Springer Biographies

More information about this series at http://www.springer.com/series/13617

Gabriella Bernardi

Giovanni Domenico Cassini

A Modern Astronomer in the 17th Century

Springer

Gabriella Bernardi
Torino
Italy

ISSN 2365-0613 ISSN 2365-0621 (electronic)
Springer Biographies
ISBN 978-3-319-63467-8 ISBN 978-3-319-63468-5 (eBook)
DOI 10.1007/978-3-319-63468-5

Library of Congress Control Number: 2017950280

Front cover image: *Giovanni Domenico Cassini*, an Italian-French mathematician, astronomer, engineer, and astrologer, vintage engraved illustration. Le Magasin Pittoresque - 1874. With permission by Science Photo Library.

Back cover images: *Giovanni Domenico Cassini*, an Italian-French mathematician, astronomer, engineer, and astrologer, vintage engraved illustration. Le Magasin Pittoresque - 1874. With permission by Science Photo Library.
Cassini was launched in 1997 with the European Space Agency's (ESA) Huygens probe, Cassini was the first spacecraft to orbit Saturn. Among Cassini's objectives was the study of Saturn's rings, Titan's atmosphere, and the behavior of Saturn's magnetosphere. Cassini ended its mission with an intentional dive into Saturn's atmosphere on September 15, 2017 ***Mission Duration:*** *October 1997- September 2017*; ***Operating Network:*** *Deep Space Network*; This image was last Updated on Sept. 18, 2017 by editor *Ashley Campbell*. With permission by NASA.

Printed on acid-free paper

This Springer imprint is published by Springer Nature
The registered company is Springer International Publishing AG
The registered company address is: Gewerbestrasse 11, 6330 Cham, Switzerland

To Alberto

Preface

When I was a young physics student at the university, the name of the astronomer Giovanni Domenico Cassini evoked a central street in Turin, my town of birth, and a well-known meridian in Bologna. So, I remember some Italian professors assumed that he was from Bologna.

Recently, the name of Cassini is back in the spotlight thanks to the NASA space mission Cassini–Huygens that bears his name, but in the literature, little content has been devoted to his life, only a few articles or some brochures.

Several years after my graduation, some scientific journalists told me about Cassini and his origin, so I discovered a pretty village in Liguria, and I went out to compose an "astrotourism" feature on Perinaldo (never paid, the magazine closed thereafter). There I learned that Cassini was born probably in his family castle, and one of his nephews left this house and joined him in Paris to become his assistant astronomer.

Today, the Perinaldo's castle is inhabited and not open to the public, but the tourist can visit the Cassini Museum, the "planet" street, and some astronomical garden and a permanent scientific exposition. On some nights, it's possible to visit the little domed astronomical observatory above the Town Hall.

Here began my idea to write a book about this great astronomer of the past; but despite this, little is known about his life in Italy or elsewhere in the world.

I wish to thank my friend and colleague scientific journalist Giuditta Bricchi for having informed me about the importance of Perinaldo in the life of Cassini. I also thank Marina Muzi for her availability during my visit to Perinaldo and its museum.

I am grateful to the Springer staff, in particular to Marina Forlizzi for having set me on the right path, as executive editor, and to Barbara Amorese and Isabella Mantovani for their professional support. Thanks also to the copy editor Ken Quinn for having improved the manuscript.

Finally, I wish to express my gratitude to my husband for his constant support and for having provided the maps in this book.

Torino, Italy Gabriella Bernardi

Contents

Chapter 1
Introduction

Giovanni Domenico Cassini, or Cassini I, is the complete name of the famous astronomer of the 16th century who founded a dynasty of scientists including astronomers, cartographers, and ending with a lone biologist. This remarkable family scored the extraordinary record of a total of six astronomers, four Cassinis and two nephews belonging to the Maraldi lineage, who worked at the Paris Observatory. Yet, the most famous of them remains Cassini I, whose fame is mostly linked to the rings and the satellites of Saturn, as well as to his earlier realization of the great meridian line in the Basilica of San Petronio in Bologna.

Was Giovanni Domenico Cassini from Bologna then?

No, only in 1702, ten years before his death, the city of Bologna conferred on him its citizenship with a decree of the Senate, fulfilling one of his dearest wishes.

So was he French?

No, but Cassini I obtained French citizenship in 1673 from Louis XIV, and he married a French noblewoman.

Already a French citizen for more than 20 years, at the age of seventy he went to Italy with his second son Jacques Cassini, or Cassini II. On this journey, he visited Bologna to accommodate the above mentioned great meridian line in the Basilica of San Petronio, and later he visited for the last time a village not too far from Genoa, where he was born in his family's castle on the 8th of June, 1625. The name of this small hill-top town is Perinaldo, and our voyage begins right there in order to the discover the first real European astronomer in the history of science.

What did Giovanni Domenico Cassini looked like?

In his portrait in Bologna's Rectorate, Cassini looks like a middle-aged man: not too young, but not so old. Actually, there are many portraits of him made at various ages, but in my opinion this one, which belonged to Cardinal Filippo Maria Monti, reflects his essence (Fig. 1.1).

Cassini wears stylish clothes such as a silk cloak with a sober, lace, white collar and carries a large dark curled wig as was fashionable at that time. But the focus is on his face.

G. Bernardi, *Giovanni Domenico Cassini*, Springer Biographies,
DOI 10.1007/978-3-319-63468-5_1

Fig. 1.1 Portrait of Giovanni Domenico Cassini

The face brightens the whole painting, showing his regular features and especially his eye-catching gaze. The man who is watching a bystander is a wealthy man of the 17th century, like many in other similar portraits of that period, but one detail stands out: the left eye brow and the eye itself are surrounded by wrinkles, and it seems he almost wants to wink at you, if it weren't for the serious attitude of the pose.

These might even be an indication of professional stress, consistent with a habitual squeeze to close the left eye lid in order to observe through the telescope with the right eye. Here is Giovanni Domenico Cassini as I like to imagine him. A serious and professional man, with an indefinable twist suggesting intelligence and passion.

Part I
Liguria

Liguria is the name of a small Italian region, enclosed between hills and sea. It is renowned for flowers, olive trees, and Giovanni Domenico Cassini, who was born in a small town on the top of one of its hills (Fig. I.1).

His education began at home, and then he was sent to the Jesuit College in Genoa. After completing his studies, he started to earn a living by teaching mathematics to the sons of the noble Genoese families.

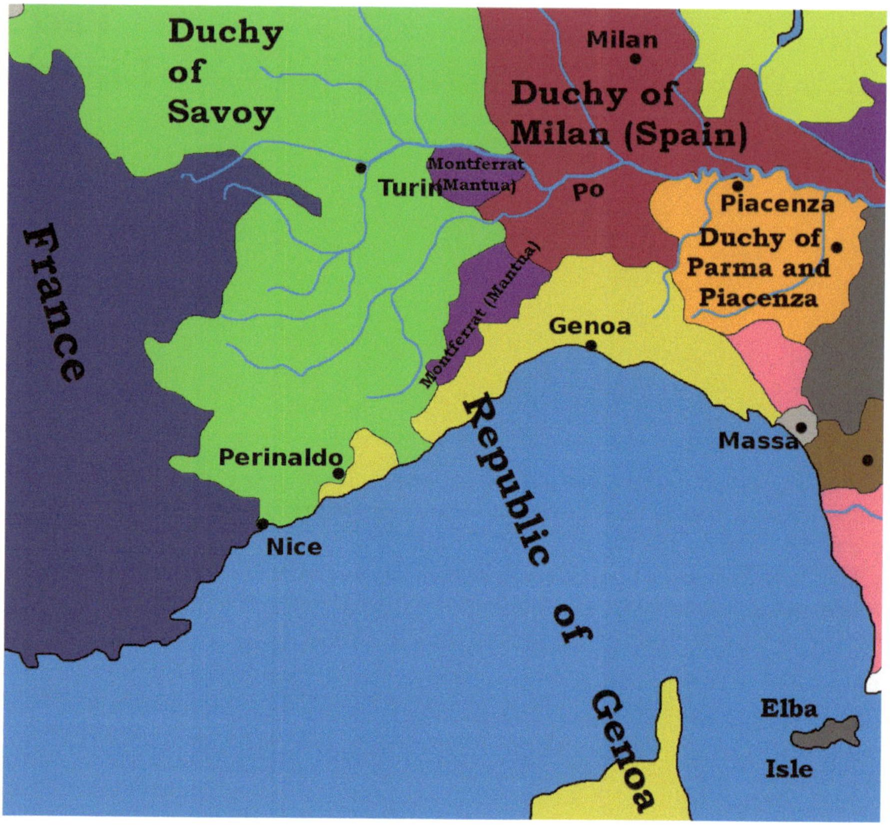

Fig. I.1 Map of northwest Italy in 1700. Perinaldo was on the border between the Republic of Genoa and the Duchy of Savoy

Chapter 2
Perinaldo, Land of Astronomers

Je suis né le 8 juin de l'année 1625, et non en 1623, comme le prétend l'abbé Giustiniani dans son ouvrage degli Scrittori Liguri...

Mémoires pour servir à l'histoire des sciences et à celle de l'Observatoire de Paris, Parigi, Bleuet, 1810
Jean-Dominique Cassini

"I was born on June 8 of the year 1625, and not in 1623, as claimed by the Abbot Giustiniani in his book on Ligurian writers..."

The astronomer Giovanni Domenico Cassini opens his autobiography with these lines in French that specify his exact birthdate. Cassini on this occasion wants to emphasize and correct the error reported by the Abbot Michele Giustiniani in his work, *Ligurian writers*, a book collecting various information about the most important representatives in the fields of Letter and Sciences from Liguria, the region in the Northwest of Italy, where he was born. The attention to such details shown by Cassini should not surprise the reader. Scientific accuracy and rigor always characterized his astronomical work, and sometimes was at the core of conflicts or misunderstandings, even, for example, with the King of France, Louis XIV.

It is therefore again not surprising that the archives of the San Nicolò Parish Church of Perinaldo, the northern Italian hometown of the famous astronomer, prove him right. The Book of Baptisms of the year 1625 records Cassini's christening two days after his birth in this way:

> Alli 10 di d.° Gio. Dominico figlio di Giac. Cassino e di Tullia sua moglie è stato battezato da me sud.° Ber.do curato, tenendolo m. Ant.° Maria Crovese e Battista Cassini q. Antonio, natto dui giorni inanti.

Its English translation from the XVII-century Italian is:

> On the tenth day of this month of June, Giovanni Dominico, son of Giacomo Cassino and of Tullia, his wife, born two days ago, has been baptized by me, Curate Bernardo, with godparents Antonio Maria Crovese and Battista Cassini.

G. Bernardi, *Giovanni Domenico Cassini*, Springer Biographies,
DOI 10.1007/978-3-319-63468-5_2

Cassini's parents, Giacomo Cassino and Tullia Crovese, daughter of a public notary, had nine children in total, of whom survived only the firstborn Giovanni Domenico and two sisters Francesca and Angela Caterina. Cassini in his biography does not report any news of his parents, and only his great-grandson Jean Dominique Cassini, or Cassini IV, wrote in his papers that the Cassini family originated in Siena, in the Italian region of Tuscany.

Probably, Cassini was born in the family's manor, now known as Maraldi's Castle, but we have no documentation about the exact birthplace, so maybe it could have been in another house of the "Contrada Grande", the oldest part of Perinaldo. Today, if you stroll through the old and narrow streets, you can see in Maraldi Street a gravestone on the doorway of the castle. It is written in Italian and commemorates the three astronomers born in this house. Besides Giovanni Domenico Cassini, in fact, his nephew, son of his sister Angela Caterina Cassini, Giacomo Filippo Maraldi was born here. And the third is Gian Domenico Maraldi, the nephew of the latter, whom we will meet toward the end of this book.

Perinaldo is a pleasant place in the hills on the western side of Liguria, from which you have a spectacular view of the Ligurian sea, a part of the Mediterranean, and of the Maritime Alps in the opposite direction. Sitting on a ridge between the valleys of the Rio Merdanzo, a tributary of the river Nervia, and of the river Verbone, Perinaldo is close to the coastal cities of Bordighera and Ventimiglia, a few kilometers inside Italy from the present border with France. These places evoke quite vividly the history of the interplay between the two former states that occupied the Italian territories of Liguria and Piedmont.

Indeed, it has to be remembered that Italy became a unified nation only in 1861. Before this date, and since the fall of the Western Roman Empire in 476 AD, the Italian peninsula was divided into many small countries, which were frequently engaged in ferocious fights and wars. In the 17th century, one of these states was the Duchy of Savoy, which comprised approximately the majority of the current Italian regions of Aosta Valley and Piedmont (actually the western part of the latter) and some territories now belonging to France, like the Savoy and Nice. The Republic of Genoa, in contrast, occupied the coastal region of Liguria and the isle of Corsica. Perinaldo was a territory in contention at the border between these two states, and, at the time of Cassini's birth, it belonged to the Duchy of Savoy as part of the province of Nizza, now the French city of Nice.

Probably, the name Perinaldo is derived from Earl Rinaldo, or Rainaldo of Ventimiglia. Between 1045 and 1055, he bought and founded a castle on the site called Villa Junchi, known later as Podii Rainaldo, which in Latin means the Hill of Rainaldo. A small town grew up around this castle, passing through a series of local ownerships. In 1524, the last of the feudatory lords of the territories that included this town, Bartolomeo Doria, was forced to cede Perinaldo to Duke Charles III of Savoy as part of a dispute between powerful local families, including the Grimaldi of Monaco. Such a fight was in turn an episode of the long series of wars fought between France and Spain at the beginning of the 16th century. The duke reunited administratively Perinaldo with the County of Nice, under which it remained until

1818, reaching its present arrangement some time later when the King of Sardinia, Vittorio Emanuele II, integrated the city into the province of Oneglia (Imperia).

The rivalry between Piedmont and the Republic of Genoa over the city of Perinaldo was still burning when Cassini was growing up, under the protection of his uncle Antonio Maria Crovese, a brother of his mother and probably the same referred to as Godfather in the baptism act, who also took care of his initial education. He was a bright child, as it is attested in the writings of astronomer Valperga di Caluso from Turin: "*...sin dalla fanciullezza diede segni di gran talento...*" (...since childhood gave signs of great talent...).

Cassini's tutor was no longer able to follow him, so he spent two years being educated by Father Aprosio of Vallebona, and, at the age of thirteen, he was sent to Genoa by his family to receive higher education in the famous Jesuit College of that city.

Curiosity

As recognition of the importance of the Cassini family, during his military campaign in 1794, Napoleon himself and Massena, one of his generals, slept one night in Maraldi Castle. Moreover, in 1797, an administrative reform of the French government moved the capital of the canton from Dolceacqua to Perinaldo.

Fig. 2.1 A showcase in the little Museum of Astronomy in Perinaldo with a telescope of the 17th century that was probably used by Giovanni Domenico Cassini (Author's photo reproduced with the permission of the city of Perinaldo)

Today, Perinaldo is a picturesque little village of less than 1000 inhabitants in the interior of the Province of Imperia in the northwest of Liguria, situated at 600 m altitude at the foot of the Maritime Alps. It is only 14 km from the seaside, 50 km from Monaco, 22 km from San Remo, and 80 km from Nice.

The city hosts a small astronomical observatory and a museum (Fig. 2.1) commemorating not only the Cassini family but also the one of the Maraldis, whose family history is closely interlaced with that of the Cassinis. Its astronomical places of interests include also a "planets street", which you can walk along, a statue of Giovanni Domenico Cassini, the darkroom sundial in the Sanctuary of the Visitation, a Stars terrace, and two Arc-Meridians.

Chapter 3
Genoa and the College of the Jesuits

...Di questo tempo[1645] tra nostri scolari
esterni contavasi il Celebre Matematico Gio Dom ^{co} Cassini...
Annue memorie del Collegio di Genova...1686
by Father Nicolò Gentile

As reported in the Italian version above, Nicolò Gentile, a Jesuit priest in Genoa, wrote in his memoirs about this college: "...At the time [1645], present among our non-residential students was the famous mathematician Gio Dom ^{co} Cassini..."

This is a precious witnessing of his time at the college, because Cassini in his memoirs does not report the exact dates of his stay, but, also from this record, it can be assumed that his attendance occurred from 1638 to 1646. The young Cassini left Perinaldo to continue his studies in Genoa, in one of the most prestigious Italian Jesuit colleges (today's University building, Fig. 3.1), following the so-called *Ratio Studiorum*. His astronomical career outside of Perinaldo would have started soon afterwards.

The *Ratio Studiorum*, from the Latin "study plan", was a particular teaching method followed within the Jesuit community as the result of a long-standing development. It was characterized by a gradual teaching procedure, which tried to favor a steady and natural learning, and included written examinations and awards. The basic course of five years was of a humanistic leaning. The classic Latin authors (except for Cicero) were studied, obviously in the original versions, and the students had always to speak in Latin, with the exception of recreation time. The best students also pursued additional courses that eventually, in the language of the time, would lead them "*ad perfectam eloquentiam*" (to perfect eloquence). Rhetoric and a daily exercise in Latin composition in prose and verse were part of this special plan, and Cassini, for a lifetime, would commemorate various events through Latin verses.

There existed several bodies of students, but they were somewhat different from present-day fraternities. For example, they were not created and managed by the students independently, but rather by the college itself, and they were reserved only for the best and the most interested students.

G. Bernardi, *Giovanni Domenico Cassini*, Springer Biographies,
DOI 10.1007/978-3-319-63468-5_3

The institution had an important role in the society, and the memories of its social and official events, such as the opening of the school year, the establishment of academies, the Master's degree, the defense of the thesis, remained vivid in Cassini's mind until the end of his life. As already mentioned, these events had significant social importance and were participated in by the main civil and religious authorities, as well as by the aristocracy. They also included cultural disputes or debates which, however, were managed almost as theatrical performances within a scenic design and with musical accompaniment. In one of these debates, in 1646, the eighteen-year-old Cassini, contributed to a discussion on philosophical and theological themes, in the presence of Cardinal Durazzo, Archbishop of Genova, and of alumni of the same college.

The years spent in Genova greatly contributed to both his knowledge and his connections, thanks to the reputation he acquired as a clever and gifted student. The well-established friendships and the attendance to optional math lessons, when he was still at the Jesuit College, became fundamental during a later stay at the Abbey of San Fruttuoso, as a guest of the Abbot Doria (a member of one of the most powerful families of Genoa). Here, for the first time, he could read and study some books that would be essential to his future profession: Euclid's *Elements*, for geometry, and the *Alfonsine tables*, or the revised *Rudolphine tables*, that is the tables of planetary ephemeris of Kepler published in 1627 and based on the precise observations of the Danish astronomer Tycho Brahe.

Before his studies in astronomy, however, the young Cassini experienced a close and soon-to-be-dismissed encounter with astrology. It all started during his stay with a friend, a Genoese patrician named Francesco Maria Imperiali-Lercaro. In his host's abode, close to the border with Lombardy, Cassini, in his memories, remembers having met a cleric coming from the island of Corsica, then still part of the Republic of Genoa. The name of the man is omitted, but it is mentioned that he owned several texts of astrology. Shortly after, the young scholar, who at the time was just 21 years old, indulged and enjoyed in making predictions, but soon his activity in this sphere ended. The reasons were both religious and scientific. Indeed, astrology was widely practiced and taught, but such practice was officially condemned by the church. Moreover, in the same period, Cassini read the treatise, "*Disputationes adversus astrologiam divinatricem*" [Discussion against astrology divination] by Pico della Mirandola. In this book, published in 1494, the Italian philosopher publicly condemns astrology from philosophical and religious standpoints. Apparently, the future astronomer was so much struck by this reading that he burned his predictions, and eventually his interest in astrology was quickly replaced by that of astronomy.

Cassini, after finishing the humanities, dedicated himself completely to mathematics and astronomy, having Giovanni Battista Baliani, mathematician and physicist, as his first teacher. Baliani was a politician who, at the behest of his father, served the Republic of Genoa as a senator and as governor of Savona, one of its provinces, but his passion was science.

Noticeably, Baliani was a "Galilean" scientist, although not as a student of the famous Pisan physicist. Indeed, the two scholars shared some theories and made similar experiments more or less in the same period, in particular about what we now call dynamics. His masterpiece, in fact, was "De motu naturali gravium solidorum" [On the natural movement of solid bodies] published in 1638, and later, in a revised form in 1646, where he proposed a theory about the motion of the bodies under the influence of gravity with many statements in common with that of the "Discorsi" by Galilei. Even if, eventually, this caused a debate about the scientific paternity of the theory, it is important to stress that the two scientists agreed on the refutation of the Aristotelean physics and on an experimental approach, which probably had some non-negligible influence on the scientific formation of the young Cassini.

Another very important influence of Baliani was about astronomy since, as Cassini recalls in his memoires: "*Il me fit voir un sextant astronomique que Tycho-Brahé avait fait faire pour Magini, par un ouvrier qu'il lui envoya exprès de Danemarck. Cet ouvrier ne fut pas plutôt parti que Magini vendit l'instrument*" [He showed me an astronomical sextant that Tycho Brahe had had made for Magini by an artisan expressly sent from Denmark. As soon as the artisan had left, Magini sold the instrument]. Tycho Brahe is the famous Danish astronomer, and Giovanni Antonio Magini was an Italian professor of astronomy who taught in Bologna, and it seems that this sextant is actually in the Royal Academy of Science, in Stockholm. It therefore seems confirmed that Cassini was seriously initiated into astronomy in this period, and his growing fame, going beyond Genoa and his native country, eventually brought him to the University of Bologna. However, this happened because of another personality: Cornelio Malvasia (1603–1664).

Curiosity

Like Cassini's mentor, Giovanni Battista Baliani, Marquis Malvasia also was a personality who divided his time among politics, science, military engagements, and inventions. Senator of the city of Bologna and then general of artillery at the service of the Duke of Modena, he was known for having hosted scholars in his castle in Panzano. Mathematicians, in particular, enjoyed this possibility, because he was an expert in the field. He also invented a micrometer for spyglasses, built with a set of parallel silver threads. Thanks to its good offices, Cassini left Genoa for Bologna, and it seems that all this happened because of astrology.

The facts are reported by Cassini in his memoirs, in the form of a curious anecdote regarding Ottaviano Sauli who, in 1649, was leading the troops of Pope Innocent X in an expedition against the Duke of Parma. Cassini and some friends of Sauli were chatting about these military events and, using the exact words of the astronomer: "*Les amis des Sauli m'ayant demandé ce que je pensais du succès de sa commission, je répondis ce qui me parut pour lors le plus vraisemblable, que Sauli serait vainquer*" [Sauli's friends asked me what I thought of the success of his assignment. I answered what seemed to me most likely, that Sauli would have won].

Then, this apparently innocent answer about the outcome of the battle took an unexpected twist when it was repeated to Sauli: "*Ce general, instruit et flatté de cette réponse, pensant d'ailleurs qu'elle était fondée sur des connaissances astrologiques, imagina, pour me render service, de parler de moi très-avantageusement à Bologne, et sur-tout au marquis Malvasia, sénateur fort attaché à l'astrologie*" [This general, informed of and flattered by this response, thinking also

Fig. 3.1 The Jesuits in Genoa settled near the old Church of San Girolamo Del Rosso and enlarged their premises by buying some land on which to locate their College and schools. The building began to be used in 1640 and is now part of the University premises

that it was based on astrological knowledge, decides, to do me a favor, to talk very positively about me in Bologna, and especially to the marquis Malvasia, a Senator strongly interested in astrology]. Eventually Sauli won, and Cassini's supposed astrological divination of such event was reported in Bologna and also to the Marquis Malvasia, who was serving as a senator at that time.

Apparently, for Cassini astrology was a vessel to proceed in his astronomical career: "*Celui-ci, sur ce témoignage, devint très-empressé de me connaitre, et pria le general Sauli de m'inviter de sa part à me rendre à Bologne, en me donnant l'espérance de me faire obtenir une place dans la célèbre Université de cette ville*" [This person [Marquis Malvasia], on this testimony, became very eager to know me, and asked general Sauli to invite me to Bologna on his behalf, giving me the hope to get a seat in the famous University of this city"].

Before Bologna, as will be explained in the next chapter, Cassini reached the Marquis in his Castle in Panzano, where the nobleman was building his observatory.

Part II
Bologna

In 1649, Cassini leaves Genoa and moves to the Castle of Panzano, at the invitation of its owner, the Marquis Cornelio Malvasia. Presently known also as Castello Malvasia, the name of Panzano refers to a suburb of Castelfranco Emilia, a small city near Modena, and thus has not to be mistaken for the city of Panzano near Florence.

About in the same period, the marquis was building his new astronomical observatory in this place, where Cassini could deepen his knowledge on the subject and make some observations himself. This observatory hosted also other astronomers, and even the Duke of Modena, Francesco I of Este, wanted to make observations from this site.

Later, encouraged and supported by Malvasia, Cassini sent his curriculum to the University of Bologna, applying for the position of professor of astronomy, where he would move in 1650 or 1651 (Fig. II.1).

Fig. II.1 Map of Bologna by Blaeu 1640

Chapter 4
Castle of Panzano

Le sénateur Malvasia était à la Villa di Pausano proche Modène, où il faisant construire un Observatoire qui devait être garni de plusieurs instrumens et orné d'une grande quantité de livres d'astronomie...

Mémoires pour servir à l'histoire des sciences et à celle de l'Observatoire de Paris, Parigi, Bleuet, 1810
by Jean-Dominique Cassini

Giovanni Domenico Cassini accepted the invitation of the marquis Malvasia to move in his castle. At the age of 24 years old, he thus left permanently his native land. As reported in his memoirs, he arrived at the castle of Panzano when: "*Senator Malvasia was in Villa Pausano, close to Modena, where he was building an Observatory which had to be equipped with several instruments and with a lot of astronomy books*".

According to the same memoirs of Cassini, it seems that the possibility to study medicine and other sciences was decisive to his acceptance of the invitation of the marquis: "*L'envie d'apprendre quelques autres parties des sciences qu'on n'enseignaint point a Gênes, et particulièrement la médecine, dont il y avait de savans professeurs à Bologne, me fit accepter avec joie la proposition du marquis Malvasia*" [The desire to learn some other parts of science, which had not been taught at Genoa, and especially medicine, in which Bologna had some learned professors, made me accept with joy the proposal of the Marquis Malvasia].

The castle of Panzano, an old medieval castle, consisted of a big palace with two high towers and a third one which housed the observatory, a garden, a big basement, the stables, a theatre, and two large courtyards. Today, it still is a private dwelling, which houses a collection of vintage cars, but the famous tower used for the observatory does not exist anymore, because it collapsed in 1899 under the excessive weight of sacks of wheat.

In this place, isolated in the countryside, Cassini continued his studies in astronomy, made observations from the astronomical tower, with the goal of updating some astronomical tables, and refined the observation instruments. A letter written in 1650 from Malvasia to the Earl Carlo Antonio Manzini, an intellectual,

G. Bernardi, *Giovanni Domenico Cassini*, Springer Biographies,
DOI 10.1007/978-3-319-63468-5_4

who had published a work about the magnetic declination needle, informs the recipient that Cassini is "his fellow student" and that he is going to explain to him the improvements he had made to the observational instruments.

But Marquis Malvasia, at the same time, played a second and possibly more important role as the patron and sponsor of his young protégé within the Bolognese society and the scholars of the university. They had to await Cassini's next birthday, when he would become twenty-five years old, because this age was the minimum to admit an applicant to the University of Bologna. Meanwhile, thanks to the influence of his sponsor, Cassini could already take his first steps in the right direction within this community (Fig. 4.1).

In this period, many important figures are remembered in Cassini's memoirs, such as his mathematics and astronomy teachers in Bologna. Among them, we can find Ovidio Montalbani, Pietro Mengoli, and Earl Carlo Antonio Manzini, along with the Jesuit fathers Gianbattista Riccioli, Lorenzo Grimaldi, Bettini, and Father Giovanni Ricci, professor of mathematics at the University. The latter, in particular, had been a pupil of Bonaventura Cavalieri, who had published an astronomical

Fig. 4.1 Portrait of Cornelio Malvasia

work with tables entitled *Directorium Generale Uranometricum,* and whose chair would be taken by Cassini four years after his teacher's death.

Curiosity

In the memoirs of Cassini regarding the period spent at the Castle of Panzano, we can read a curious passage, which is quite useful to understand the personality of the young scientist: "*Il* [marquis Malvasia] *avait coutume de faire imprimer tous les ans un Journal astrologique dont il faisait present à ses amis; Je* [Cassini] *lui représentai qu'il serait plus honorable de calculer d'après les éphèmerides des tables astronomiques plus modernes, et de laisser à part les predictions astrologiques qui n'avaint aucun foundement solide*" (He [Marquis Malvasia] used to print every year an astrological journal which he then gifted to his friends. I [Cassini] pointed out to him that it would have been more honorable to calculate the most modern tables of astronomical ephemerides, leaving aside the astrological predictions, which have not no solid basis).

These lines indicate that astrology was practiced among the intellectuals of the time, but Cassini was not afraid to express his opinion, even when it was contrary to that of the majority, and above all to that of his patron. Maybe it was just because of his young age, or he was lucky, and the Marquis was not touchy. We do not know, but this remained a trait of his personality for the rest of his life. He would not be afraid to state frankly his professional opinion to anyone, including emperors.

During his life Cassini tried to fight astrology. In his opinion, not only did this discipline have no scientific basis, but also its alleged predictive power was proven wrong by facts. As he reports at the end of this anecdote, in this case, he prevailed and managed to convince the Marquis: "*Ce bon conseil que je lui donnais fut bientôt confirmé par un évènement assez singulier qui lui fit reconnaitre, que ce n'était que par hazard que le predictions astrologiques avaient quelques succès. Il avait prédíti dans son Almanach una grande tempéste pour un certain jour, et ce même jour un ouragane et une grêle furieuse ruinèrent les campagne d'aleutour; le marquis Malvasia vint me trouver son livre à la main pour me convaincre de la joustesse da sa prediction. Fort bien, lui répondis-je; mais voyons un peu sur quel fondement vous vous être appuyé et repassons les calculs. Ce qui fut fait assitôt. Mais il se trouva, à ma grande satisfaction, que c'était par une faute d'impression que l'ou avait marqué dans les éphèmerides une configuration qui n'avait point en lieu, et d'après laquelle le sénateur avait conclu l'événement de la tempête, qui n'aurait pas dû avoir lieu si les éphèmerides eussent été justes. De ce moment Malvasia prit le parti de calculer lui-même de nouvelles éphèmerides*" [This good advice that I gave him was soon confirmed by a singular event that made him recognize that it was only by chance that the astrological predictions had some success. His almanac had foretold a great storm for a certain day, and on that very day a hurricane and a furious hailstorm devastated the surrounding farmland. The Marquis Malvasia came to me with his book to convince me of the precision if this prediction. 'Well,'—I replied—'but let us see on which basis you came to these conclusions, and let us review the calculations.' This was done immediately. But he found himself, to my great satisfaction, that, because of a misprint in the ephemeris, he had used a configuration that was not possible, and according to which the senator had concluded the event of a storm that would not have happened if the ephemeris had been true. From that moment Malvasia, resolved to calculate himself the new ephemeris].

Chapter 5
Archiginnasio

Conductio D. Doctori Jo. Dominici de Cassinis
[Hiring of Doctor Jo. Domenico Cassini]
Archivio Notarile—Ramponiesi Paride 6-3-4,
1647–1651. Bologna, Archivio di Stato

The "Studium" (study) or university headquarters was the Archiginnasio palace, very close to the Basilica of San Petronio, the main church of Bologna. It was built there by Pope Pius IV, at his own expense, in 1563, to host teachers and pupils. The choice of the location, just 12 m from the Basilica, was not accidental because he did not want San Petronio to become larger than St. Peter's, in Rome, so he simply put a physical barrier to stop the growth of the rival building.

From documents found in the state archives in Bologna, Cassini presented to the Senate, on March 4, 1651, his application for the Chair of Astronomy. The date seems to be in contradiction with the accepted one of 1650 reported by the Ligurian scientist in his memoirs, but it is possible that the latter simply referred to the academic year of 1650–51. This position was nothing less than the "Cavalieri" chair which, as mentioned in the previous chapter, had remained vacant since the death of Father Bonaventura Cavalieri, in 1647. It was quite a demanding burden for such a young scientist. Cavalieri, in fact, was not only an astronomer but especially also a mathematician, who had arrived in Bologna in 1629 under the sponsorship of Galileo Galilei. His celebrity in the history of science is linked to his pioneering studies on geometrical methods for the calculation of areas and volumes which are considered as forerunners of the present integral calculus, published in the book "*Geometria indivisibilibus continuorum nova quadam ratione promota*", and known under the name of "Geometry of the indivisibles" (Fig. 5.1).

The application consists of a sheet of paper written by Cassini himself, which includes also his résumé: for five years, he taught philosophy and mathematics to the descendants of the Genoese nobility. On the back of the paper, there is an annotation by the selection board in which Cassini is characterized as: "*very virtuous person*", educated in the profession and worthy to hold the chair. On April 12, 1651, with a notary deed from Paride Ramponiesi, the Senate of Bologna, under the pontificate of Innocenzo X, named Cassini as the Chair of Astronomy for five years.

G. Bernardi, *Giovanni Domenico Cassini*, Springer Biographies,
DOI 10.1007/978-3-319-63468-5_5

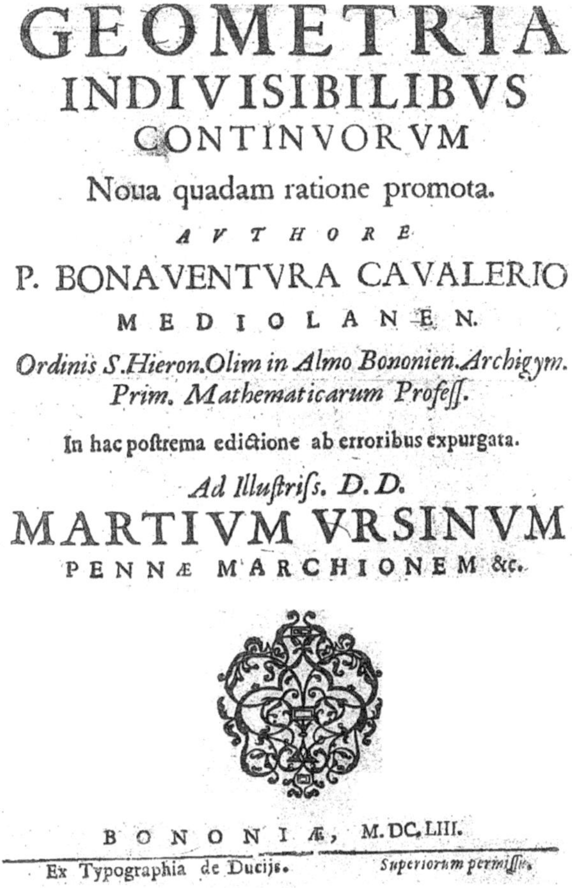
GEOMETRIA
INDIVISIBILIBVS
CONTINVORVM
Noua quadam ratione promota.
AVTHORE
P. BONAVENTVRA CAVALERIO
MEDIOLANEN.
Ordinis S.Hieron.Olim in Almo Bononien.Archigym. Prim. Mathematicarum Profess.
In hac postrema editione ab erroribus expurgata.
Ad Illustriss. D.D.
MARTIVM VRSINVM
PENNÆ MARCHIONEM &c.
BONONIÆ, M.DC.LIII.
Ex Typographia de Ducijs. *Superiorum permissu.*

Fig. 5.1 Frontispiece of Bonaventura Cavalieri's "Geometria Indivisibilibus"

His teaching duties consisted of giving lessons "*at the fourth hour of the evening*" or, in other words, during the afternoon classes. His course, named *Theory of the Planets*, constituted part of the study plan called "*Ad Mathematicam*", namely *To the Mathematics*, which included a course of *Euclidean geometry*, assigned to Father Giovanni Ricci, who also taught in the afternoon.

As pupil of Galilei, and despite his mathematical inclination, Cavalieri had contributed to disseminating the new principles of the experimental science in a quite conservative environment, as the University of Bologna was at the time, which was damaging the reputation of this historical institution. These state of affairs were apparent at the time of Cassini. Indeed, the University of Bologna was important in Italy and in the Europe, but not so much as in the past. Decadence was in progress in spite of the measures undertaken by the authorities to restore the professionalism of the university.

The academic environment of the city, however, was anything but poor. Alongside the university, there was the Jesuit College of Saint Lucia which featured an efficient educational system. This school played a role as a competitor of the *Studium*, but, despite this institutional rivalry and unlike in other similar situations, the two bodies enjoyed a fruitful exchange of experience, scientific data, and opinions among their members. In this case, therefore, the various different, and somewhat opposite, scientific conceptions were based on frequent and loyal relationships, from which Cassini, in particular, could take great advantage. Actually, as mentioned in the previous chapter, Cassini would meet many future colleagues from the university, like Giovanni Ricci, Pietro Mengoli, Grimaldi, and Ovidio Montalbani, but also Jesuit scientists like Bettini, Lorenzo Grimaldi, and Giovanni Battista Riccioli, and remarkable amateurs like the aforementioned Earl Carlo Antonio Manzini.

The "*Rotuli dello Studio*", which can be translated as *Rolls of the Study*, are large sheets of vellum adorned with miniatures, written up year by year, reporting the list of lectures, the subjects taught, the hours, and the regulations. The "Roll of the Year" was carried in procession on the saint day of San Petronio, the patron saint of the city who was the bishop of Bologna in the fifth century, and then remained affixed, for a few days, to the gates of the university.

So, when Giovanni Cassini was sworn into office, his name was registered in the list of teachers of the "*Rotuli dello Studio*" of that year with an annual salary of 600 *Bolognese lire*, the coin used at the time.

As per the regulations written on the "Rolls", the lecturers of the university could not refrain from teaching, and they could skip the lesson only if sick or for public-service reasons. Cassini however, as we shall see, would often be excused from his teaching duties because of other obligations required him by the Pope himself. The lecturer had to start and finish the "lettura", i.e., the lesson, at the sound of a bell. Actually, it was one of the four bells of the belfry of the San Pertronio basilica, called "*la scolara*", which in English was meant as "the bell of the school", which tolled each hour.

Curiosity

From the document authorizing the hiring of Cassini, written in Latin and preserved in the state archives in Bologna, it can be understood from where Bologna could collect the funds needed to pay the salaries of its university's teachers. Curiously, they derived from the "*Gabella grossa*", namely a sort of customs fee. More exactly, this duty was paid on all goods imported and exported by land or *by water* in Bologna. At the time, indeed, even if Bologna is not coastal city, there was an extensive network of waterways that connected the Po River (the longest in Italy) to the city, and, as we will learn in the following chapters, these would an important part in the working activity of Cassini during his years in this city (Fig. 5.2).

Again, from the "*Rotuli dello Studio*", we can learn that the astronomy teacher was required to provide each year a complete table of ephemerides of all the planets for each day of the coming year. Although such a task cannot be considered

Fig. 5.2 A glimpse of the past appearance of Bologna: one of the few remaining watercourses of the town

surprising, the motivations were quite different from what one could now expect, since they were rather based on astrological beliefs. On the first of January, indeed, the tables were delivered to the Rector, who made them available to the physicians. During those times, in fact, the medical profession was also based on alleged astronomical, or rather astrological, influences on the human body.

The dress of the lecturers is also again reported in "Rolls", and it consisted, unless the lecturer was in the clergy, of the "Toga Dottorale", i.e., the Doctoral Gown with wide sleeves. This specific dress had the purpose of making the members of the university distinguishable from all the others, who had tight sleeves instead.

It is interesting to know that father Giovanni Battista Riccioli, a Jesuit colleague in Bologna, attempted to convince Cassini to take vows and join the Society of Jesus, but he always refused.

Chapter 6
Comet 1652, the First

...insomni studio
per gelidas noctes...

[...with sleepless study
in the chill of the nights...]

De Cometa [The comet] *by Giovanni Cassini*

In 1652, just before Christmas, a comet was noticed approaching the Earth. From the city of Bologna, it was visible at the zenith, and it was observed also by the archbishop of the city. As reported by Cassini in his memoirs, this circumstance required again his presence at the castle of Panzano, by request of Malvasia: "*...le marquis Malvasia voulut absolument que je me transportasse avec lui et Beringelli Geri mon disciple à sa maison di Pansano...*" [...the Marquis Malvasia absolutely wanted that I come with him and my student Beringelli Geri to his home of Panzano...].

Cassini thus relocated to the Panzano Castle, far from the city, which actually hosted a very well-equipped observatory that would enable him to make better observations. One can appreciate the sophistication of this observatory by the quote of Cassini, who denoted this place as *an Italian Uraniborg*, namely the castle of the Danish astronomer Tycho Brahe, which about 60 years before had become famous as the most advanced observatory of the time. Although one might admit to a bit of exaggeration towards his friend and patron Malvasia from the Italian astronomer, this judgment from a professional, who had already shown little reluctance in expressing his opinion and who also contributed to the realization of the observatory, cannot surely be underestimated. In this observatory, he was also able to use new tools built especially for the occasion. His plan was to follow the comet's movement night after night and to determining its latitude and longitude. Duke Francis of Modena, a curious man interested in astronomy, was among the guests participating in some of the observations, and Cassini dedicated the work that was completed after his observations to him: "*De Cometa anni 1652 et 1653*" [The comet of the year 1652 and 1653].

Indeed, Cassini crafted a remarkable work, and in this publication he gave detailed observational data, which included drawing of the comet's positions with

G. Bernardi, *Giovanni Domenico Cassini*, Springer Biographies,
DOI 10.1007/978-3-319-63468-5_6

respect to the constellations. This work, written in Latin, also included the author's personal considerations about comets, an argument that he will study for a lifetime.

As an example of such considerations and of their importance in the scientific debate of the time, it can be mentioned that Cassini stressed that the comet, which passed at the zenith and at the end of his observations was above Saturn, had no significant parallax. This, in other words, meant that it was quite far from the Earth.

At the time, the most common hypothesis on the origin of comets was that these objects were "*exhalaisons*" or emissions of the Earth. This was based on Aristotelean physics, which divided the Universe into the celestial, incorruptible world, and the terrestrial one, the place of transient and variable phenomena. Comets, in this sense had to belong to this "lower" world.

Cassini, in his memoirs, reported that at first, "*...je ne m'éloignais guère de l'hypothèse la plus commune sur la générations des comètes, avec cette difference que j'attribuais leur origine au concours des exhalaisons tant de la terre que des astres; car je supposais que chaque astre a une atmosphère qui s'étend fort loin, et qui se mele avec les atmosphère des autres astres*" [I did not distance myself very much from the most common assumption on the generation of comets, with the difference that I attributed their origin to the concurrent emissions of both Earth and the stars; this was because I supposed that each star had an atmosphere that extended very far, and which interacted with the atmosphere of the other stars].

But afterwards, comparing the observations of several comets, Cassini noted that their movements seemed uneven and "*...je reconnus qu'il pouvait se réduire à l'égalité sur une ligne circulaire fort excentrique à la terre*;..." [I realized that the cometary movement could be led back to equality on a very eccentric circular line to the Earth]. In more modern language, they could be interpreted with a common model as objects having very elongated orbits, that is, with a large eccentricity.

These considerations made the astronomer speculate that the comet seemed very reasonably to match the ancient hypothesis of Apollonius Myndien, in fact "*...et ayant vu dans les dernières observations cette comète passer par le zenith et n'avoir point de parallaxe sensible, j'estimai fort raisonnable l'hypothèse ancienne d'*Apollonius Myndien,..." As the Italian astronomers explained immediately afterwards, according to this hypothesis, that the comets were "perpetual stars" whose movement is so eccentric with respect to the Earth that they are visible only when they are approaching their perigee.

The apparatus used for observing night after night the comet is described and depicted in his publication. It was a large instrument made of wood and with the shape of a compass. The angle between the two legs was adjusted in order to make each of them point to one of a pair of stars. Then, the angle between these two objects was estimated thanks to a series of mobile and fixed parts connected by screws. This particular tool appears in Cassini's publication in response to the expressed wish of the Marquis Malvasia.

Curiosity

"*De Cometa anni 1652 et 1653*", as already mentioned, was the first scientific study written by Cassini, and it is also the only one reporting the Ligurian origin of the scientist, who is denoted as "*Jo. Dominicus Cassinus genuensis In Bononiensi Archigymnasio publicus Astronomiae Professor*" [Gio. Domenico Cassini, from Genoa, public Professor of Astronomy at the Archigymnasio of Bologna].

Actually, the genesis of this work is quite peculiar. It stems from the competition for the paternity of the scientific discovery, and possibly, from the academic rivalry between the university and the Jesuits.

Marquis Malvasia immediately gathered in his castle some printers from Modena, so that the tables could be printed as soon as the data and the drawings were collected. This provided a decisive advantage over the competitors with respect to the publication of the results.

Indeed, in the introduction Cassini, reports that the comet appeared on December 20th, at the fourth hour of the night, between the Pleiades and the constellation of the Taurus, close to the neck of the Bull: "*Ibi igitur die vigesima hora noctis quarta observavimus Cometam cum Pleiadibus, et cum stella proxima in collo Tauri talem officere configurationem...*" These words are dated January 6, 1653, and a very short time later appeared the printed version of the book, which was immediately sent to Prince Leopoldo de' Medici in Florence (an Italian banking family and political dynasty), who was big fan of scientific disciplines, just a few days before the arrival of a similar work by the Jesuits.

Finally, in a draft of a letter written by Cassini to an unknown recipient after the publication of his *De Cometa*, the Italian scientist emphasized the mathematical and astronomical basis of his work. In particular, he explicitly rejected any astrological considerations, and it seems that he also intervened to reassure the people who were terrified by the arrival of the comet.

Chapter 7
The Meridian Line or Heliometer in the Basilica of San Petronio

Ignace Dante, dominicain, avait entrepris
de tirer une grande ligne dans l'église
de Sainte-Pétrone pour les observations du soleil...

Mémoires pour servir à l'histoire des sciences
et à celle de l'Observatoire de Paris, Parigi, Bleuet, 1810
by Jean-Dominique Cassini

Originally, the construction of the meridian line in the Basilica of San Petronio was undertaken between 1575 and 1576 by "*Ignazio Danti, Dominican*," who "*had set out to draw a line in the Church of St. Petronius for observations of the Sun...*". This is how Cassini begins, in his memoirs, to relate his big adventure connected to this church and its sundial or, as he called it himself, heliometer, because he measured the apparent diameter of the Sun during the year.

Ignazio Danti (1536–1586) was an Italian astronomer, mathematician, and cosmographer for the Duke of Tuscany. Then he left Florence to become professor of mathematics at the University of Bologna. He built his meridian line inlaid in the paving of the left aisle of San Petronio. Already designed for the Basilica of Santa Maria Novella in Florence, eighty years before Cassini, this instrument had a very important purpose. It was built to make accurate astronomical measurements for completing the reform of the Julian Calendar, which was at the time under study, and would have brought about the transition to the currently used Gregorian one.

In order to better understand the last statement, it will be useful to recall that, until 1582, the calendar used for civil and religious purposes was the Julian calendar, which derives its name from Julius Caesar, who introduced it in 46 BC in the Roman domains. This belongs to the category of the so-called *solar calendars*, namely those based on the duration of the *tropical year*. The tropical year is the time taken by the Earth (or by the Sun, from our point of view) to complete a seasonal cycle, going for example from one equinox or one solstice to the subsequent one of the same kind. As is known, this differs from the so-called sidereal year, which is the time taken by the Earth to complete a 360° orbit with respect to the "fixed" stars, because of the precession of the equinoxes. Solar calendars are based on the tropical year exactly because of their seasonal periodicity.

G. Bernardi, *Giovanni Domenico Cassini*, Springer Biographies,
DOI 10.1007/978-3-319-63468-5_7

The latter, indeed, is fundamental for farming needs. Moreover, in the first Council of Nicaea of 325 AD, the Church linked its most important festivity, Easter, to the same cycle, by setting the rule of celebrating it on the first Sunday after the first full Moon of spring. Obviously, this requires to know the exact moment of the spring equinox and therefore the duration of the tropical year.

The Julian calendar, computed by the Hellenistic astronomer Sosigenes of Alexandria, had established a duration of 365.25 days, which eventually translated in the rule of having one leap year every four. It is well known that such duration slightly exceeds the duration of the mean tropical year, which is approximately 365.2422 days, so that the Julian calendar lags gradually behind the actual seasonal periodicity.

This was also well-known five centuries ago, when, after several years of debates, in 1582 another council decided to reform the calendar, according to the so-called *Gregorian calendar*. The meridian line of Ignazio Danti fitted exactly these needs, since it allowed the high-accuracy determination of the length of the tropical year, upon which the rules used to define the leap years depend.

Meridian lines, in fact, are a particular kind of solar clock that do not indicate the time, like a sundial. Rather they denote the instant at which the Sun crosses the local meridian. Sundials are, in general, made up of a plate, possibly with some kind of hour lines marked on it, and a stick, or a rod, called the gnomon. The gnomon casts a shadow on the plate, which changes according to the local solar time. Meridian lines, instead, are formed by a line drawn along the local meridian, i.e., along the North-South direction, in a closed room.

A hole drilled in the rooftop of the room perpendicular to the line, called the *gnomonic hole*, can thus project the image of the Sun, which will intersect the line exactly at the local noon. The point of intersection will vary, according to the day of the year, within a range determined by the two solstices. Precisely, at least in the Northern Hemisphere, at the winter solstice the image will cross the meridian at its most northern point, while at the summer solstice, this will happen at the most southern one.

Moreover, the point of the meridian line corresponding to the equinoxes can be obtained in two ways (see Fig. 7.1):

- either by taking the half-angle (bisection) between the directions from the gnomonic hole and the two solstices (which form an angle of 47°, namely twice the inclination of the ecliptic),
- or by taking the intersection between the meridian line and the perpendicular to the direction pointing to the astronomical North (the Pole P in the figure).

Finally, the time needed for the Sun to return to any point of the line from the same side is exactly one tropical year. One can say, therefore, that a meridian line does not mark the time, but rather the day of the year.

However, Danti's meridian line still suffered from a number of problems. The gnomon was not good enough, as reported by Cassini: "*...les rayons du soleil à midi allaient réncontrer les colonnes...*" that is, the rays of the Sun at midday hit the

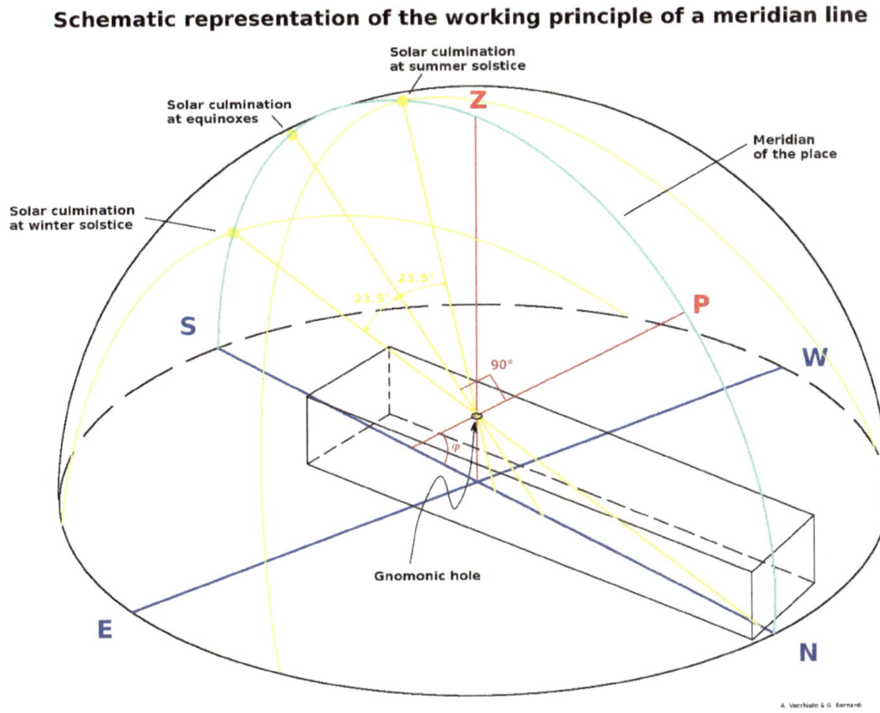

Fig. 7.1 Schematic representation of the working principle of a meridian line. The blue lines represent the cardinal directions of the place, so that the N-S one is aligned with the meridian. Z is the zenith and P is the (north) celestial pole, therefore φ is $(/2-\phi)$, where ϕ is the local latitude. See text for a more detailed explanation

columns. Moreover, the line diverted more than nine degrees from the north-south direction, namely from the meridian.

Furthermore, there were no markings on the line to infer the altitude of the Sun. In addition to these problems, at the time of Cassini, the fate of Danti's solar clock was already sealed because of the impending expansion works planned for the basilica. These, in fact, required the demolition of the wall in which the gnomonic hole had been drilled. However, the city government decided to use this circumstance as an opportunity to replace the condemned meridian line with a larger and more accurate one.

On June 12, 1655, Cassini was commissioned by the "*Fabbricieri di San Petronio*", or the Conservators of St. Petronio, to build a new meridian line, and he realized that it was possible to draw a long base line passing between the columns bases, thus avoiding the interruptions that troubled the then existing one. Danti's gnomon hole was raised, and the young Ligurian found in the vault a point to "*1000 pouces du pied de Paris*", at the time the most widely used measurement unit for high-precision instruments. Precisely, the "pied du Roi" or "Pied de Paris" (foot, in English terms) corresponds to 0.324839 m, and each foot was divided in 12 "pouces" or inches, which implies that the height of the hole was about 27.07 m. In this way, Cassini's was able to transmit the image of the Sun at noon throughout the year.

Moreover, the image projected by such a pinhole on the floor of the Basilica was not just a spot of light; it was the same image of the Sun, reversed, as in a darkroom. This is a common feature of the meridian lines, which is indeed important for the scientific exploitation of the instrument that Cassini was designing, as will be explained later.

The original project, detailed in "*De novo gnomon meridian in D. Petronij Templo construendo*", was presented to Marquis Innocenzo Facchinetti, president of the *Fabbriceria* of San Petronio, and underwent the review of several mathematicians and astronomers. Indeed, quite a number of objections were raised, only in part of a technical nature. Surely the problem was a complex one because of the architecture and the configuration of the building, and many remained skeptical despite the accurate measures presented in the project plans. Yet, there were probably other, less objective, reasons. As reported about 50 years later by the Bolognese astronomer Eustachio Manfredi, Marquis Malvasia openly supported the proposal at the Senate, thus providing an important political endorsement to Cassini. On the other hand, thanks to the same Malvasia, the latter held a prestigious professorship, despite his status as a foreigner, and his very young age, only 30 years.

It is thus reasonable to suppose that this status of affair aroused an unfavorable context around the "privileged stranger" and his project. Jealousy and envy might also have been stimulated by the fact that many mathematicians before him had tried to solve the problem, without success. Nevertheless, the scientist recalls with apparent relief in his memoirs "*...la permission de tenter mon enterprise: elle me fu enfin accordé*", that is, despite the many obstacles, he was finally granted the permission to undertake his enterprise.

In the end, to get rid of any doubt, Cassini invited his colleagues and "*tous les savans de Bologne*" (all scholars Bologna) "*pour etre témoins du succès de mes tentatives*" (to be witnesses to the success of my attempts) on June 21–22, 1655, that is during the summer solstice. The meridian worked perfectly thanks to Cassini's expertise and his care in the realization of the device. As an example of the difficulty of such an endeavor, the iron base line, which represents the path of the Sun and crosses obliquely the floor of the church, had to be laid down with extreme precision. This required many measurements and an accurate leveling of the floor, which was carried out with a water-canal system.

With such a precise instrument, Cassini was able to pursue a remarkable and very advanced research program, for the time, which extended considerably the original purpose of measuring the exact duration of the tropical year needed for the calendar reform.

First of all, as mentioned above, the angle between the solar directions at the two solstices is twice the inclination of the ecliptic. Therefore, a meridian line is a perfect instrument to estimate such a quantity, and, in fact, according to the memoirs of the astronomer, the obliquity of the ecliptic was finally estimated to 23°29′. However, the outcome was not so straightforward as it might seem at first sight.

Indeed, when observed from the ground, the incoming light rays from the Sun deviate from a purely straight line because of the refraction of the Earth's atmosphere. Thus, in order to avoid altering the estimation, the measures have to be corrected for this effect. In those times, the most accredited theory of refraction was that of Tycho Brahe, who believed that, at 45° of altitude the refraction was zero, so Cassini at first evaluated the ecliptic according to these predictions. But the value of the ecliptic inclination is used to predict the ephemerides of the celestial bodies, which can thus be used to make a verification a posteriori of the estimation.

In particular, the young astronomer found that the solar longitudes were not in agreement with the predicted values. This obviously could depend on many other factors, but eventually, after careful checks and verifications, Cassini came to the conclusion that the "culprit" was the refraction, and that Tycho's theory had to be modified, together with the estimation of the obliquity of the ecliptic. His theory of refraction remained the best one for more than a century, before being replaced in the second half of the 18th century.

In addition to these results, as Cassini reports in his memoirs, with the meridian line he was able to obtain other scientific achievements regarding the determination of the solar parallax, and the more "geodesic" determination of the relationship between the length of the sundial and the Earth's circumference. All these data were afterwards verified and published by the astronomer Jean Picard in his book "*La Mesure de la Terre*" (Measuring the Earth).

But another problem that Cassini wanted to solve with his meridian line was of a completely different nature, and it involved nothing less than the controversy between the geocentric and heliocentric world systems. This was later argued in his work *Astronomy Reformed*, but let us first hear it in his words from the *Memoirs*: "*Un des principaux usage que je fis de mes observations à la nouvelle méridianne de Sainte-Pétrone, fut de montrer par leur moyen que l'inégalitè du movement apparent du soleil ne depend pas immédiatement de son excentricité, qui est cause que son diameter apparent paraìt plus grand dans le perigee. Mes observations firent voir que le diameter apparent du soleil, qui diminue en s'éloignant du perigee, ne diminue pas à proportion comme le movement de cet astre dans l'écliptique. Kepler l'avait déjà avancé; mais les astronomes, entr'autres le père Riccioli, n'avainent pu se persuader jusqu'alors.*"

This piece needs an explanatory introduction. According to the geocentric system, not only does the Sun orbit around the Earth, but also its orbit is circular. However it was well-known that the apparent motion of the Sun, as seen from the Earth, was slower during summer and faster in winter. It was also known that the solar diameter was smaller in the former case and larger in the latter. Both these effects were qualitatively in agreement with the hypothesis that the distance of the Sun from the Earth varied throughout the year, but how to reconcile it with a perfectly circular orbit? Apparently the solution was simple: the Earth's position was slightly off-center with respect to such orbit. In this way the speed of the Sun could remain constant along a circular orbit, and both the mentioned effects could be ascribed to a pure variation of its distance.

This scenario does not change too much if we simply consider the Sun at the center, as Copernicus did, but the formulation of Kepler's laws implied a new scenario, which was incompatible with the geocentric one. The two key facts were the elliptical shape of the orbits, and the physical variation of the speed of the Earth (or of the Sun, if we consider it from the point of view of the Earth) along its orbit, which is required by the second law.

By the first law, it is clear that the distance varies, which implies a variation of the apparent speed of the Sun and of its diameter of the same purely geometrical nature of that of the geocentric model. However, when it comes to the second law, one has to admit that the speed has an *additional* decrease due to the slowing down of the physical motion of the Sun. By simple geometric considerations, in the case of uniform motion hypothesized in the geocentric (and Copernican) model(s), the decrease of the apparent speed while moving from the perigee to the apogee is just proportional to the apparent diameter of the Sun. In the Keplerian model, instead, the Sun slows down also from a physical point of view, which means that its apparent speed decreases more that its diameter.

This allows a better understanding of the above citation of Cassini. We recall in fact that the meridian line works like a darkroom, and that the projection of the Sun on the floor is not point-like, but rather a disk. From its dimension, therefore, it is possible to estimate the apparent diameter of the Sun and, if the instrument is accurate enough, to discern between the two cases.

In practice, with his observations at the meridian line of San Petronio, the Italian astronomer proved that the solar-motion difference in speed during the year is not just proportional to the change of its diameter. Therefore, the additional physical change postulated by Kepler was real and in agreement with his second law. This cannot be considered the final demonstration of the superiority of the heliocentric model, which had to wait for Newton's formulation of the theory of gravity; however, it was an important step in this direction.

Indeed, without more stringent dynamical motivations, one can still put the two bodies on the same ground, in the sense that the Earth- and Sun-centered reference systems can be considered equivalent to each other. However, the observations proved the validity of the second law of planetary motion, and thus definitely disproved one of the main statements of the geocentric model, namely that of uniform motion along circular orbits. The importance of this fact cannot be underestimated since, as mentioned in the above excerpt, this was considered false by many astronomers like the Jesuit father Riccioli.

To sum up, Cassini used the observations at the meridian line to estimate the value of the obliquity of the ecliptic, and to demonstrate experimentally the validity of Kepler's second law. In doing so, he studied the atmospheric refraction, discovering that Tycho's model had to be corrected. He also obtained significant results about the solar parallax and the determination of the Earth's circumference. In 1656, he published an essay of his observations entitled: "*Specimen Observationum Bononiensium quae novissime in D. Petronij templo ad Astronomiae novae constitutionem haberi coepere*".

Curiosity

Today, the meridian line in San Petronio is one of the largest astronomical instruments in the world, at a length 67.72 m, in a closed building. This length corresponds to the six- hundred-thousandth part of the Earth's meridian.

When finished, it was used by Cassini and other astronomers like for example the Jesuit Riccioli, for determinations of equinoxes, solstices, and the Sun's apparent diameter and declination. Father Riccioli considered this meridian line the best in Italy and in Europe.

According to the archives of the "Fabbriceria", the costs for the realization of this solar clock were around 2000 *lire*. Cassini received 500 lire, but on September 22, 1655, another 200 lire and three years later yet another 300 lire "*as an ultimate gift*".

The sunlight paintings around the hole in the ceiling are not random, but calculated mathematically by Cassini, who used a tool he invented himself, which he called "*machine parallactique*" or a parallactic machine.

The first drawing of the meridian line of San Petronio of Cassini has been lost. The astronomer gave it as a gift to Queen Christina of Sweden when she passed through Bologna in November 1655, on her journey to Rome. This woman will have an important role in the life of the Italian scientist, which will be described in a following chapter.

As mentioned above, possibly the main result obtained with the meridian line was the experimental demonstration of Kepler's second law. The fact that this did not totally disprove the geocentric model was well-known to Cassini, who prudently continued to avoid any reference to the motion of the Sun despite the accuracy of his data. This cautious attitude can have been motivated in various ways. Partly, it can be ascribed to his rigorous scientific attitude: if the experimental evidence does not allow preferring one model over the other one, then a scientist can only say that both can be used. However, the political and theological debate around this issue surely played a role in the approach of the Italian astronomer. It has to be remembered that about 20 years before this Galileo had been brought to trial and sentenced to abjuration by the Catholic Church for such a disputation. As Heilbron said in his "Churches as scientific instruments", Cassini: "*provided strong evidence in favor of a view condemned by the Holy Office, nothing would happen provided that the happy discoverer did not insist on interpreting his discovery obnoxiously*". As a Catholic in a town of the Papal States, he did not want to clash with the Church and the Jesuits astronomers who still had great power over the scientific culture in Bologna.

Chapter 8
Superintendent of Public Waters

Je fus fréquemment distrait de mes observations astronomique…
[I was frequently distracted from my astronomical observations…]

Mémoires pour servir à l'histoire des sciences et à celle de l'Observatoire de Paris, Parigi, Bleuet, 1810
by Jean-Dominique Cassini

For Cassini, 1656 was a very successful year. He was already known throughout Europe for his comet observation, and now, after publishing the results of his observations at the meridian line, he became even more famous because of his "*big astronomical machine*". In the same year the mandate for the astronomy professorship at Archiginnasio expired, but the Council renewed his chair for another three years, with a substantial salary increase, which thus reached the amount of "*1200 lire*".

Indeed, his new and increased celebrity made it advantageous for the University of Bologna to have him as a professor. He continued in his lessons at the University and his astronomical studies in San Petronio with the heliometer. In particular, as previously mentioned, he focused on the phenomenon of atmospheric refraction, which at that time was almost unknown. For his astronomical observations, Cassini used also two famous buildings, still in existence in Bologna, namely two of the many towers that even today enrich the architecture of the city.

These structures were raised during the medieval age, in a period where many families were competing for power in the town, and probably they had both defensive and prestige-assertion purposes. However, they were also ideal for astronomical research, and, in particular, for the investigation of the atmospheric refraction, because they offered an elevated viewpoint, from which the horizon was free of obstacles, and extended the sensible horizon (namely the portion of the sky that is not hidden by the Earth) below the level observable from the ground.

Both dating back to the twelfth century, one of the towers used by the young astronomer was the *Clock Tower*, and the other was the *Asinelli Tower* that, at 98 m, is the tallest of the city. Regarding the refraction experiments, as reported by

G. Bernardi, *Giovanni Domenico Cassini*, Springer Biographies,
DOI 10.1007/978-3-319-63468-5_8

Cassini himself, the sensible horizon seen from these two buildings was "*between 13 and 14 min under the ground level for the first and 19 for the second one.*" It was not unusual to exploit the towers to carry out scientific experiments—the most famous case is Galileo and tower of Pisa—but, for example, another scientist from Bologna, the already mentioned father Ricci, also used the same Asinelli Tower for experiments on falling bodies (Fig. 8.1).

Apparently, then, Cassini was in a perfect situation. He was still young, thirty-one-years old, yet already quite famous in Europe as a mathematician and astronomer. He held a prestigious chair in one of the most renowned universities of the continent. Finally, his perfect mix of observational and theoretical skills and his unstoppable scientific activity had produced many remarkable results, with the potential of further promising developments. Everything seemed to suggest a highly

Fig. 8.1 View of Bologna and its surroundings from the top of the Asinelli tower, where Cassini and other scientists made astronomical observations and experiments

successful future for him in the field of astronomy, full of researches and discoveries. However, in this period, he could not dedicate himself to this task with the energy he would have wished. Indeed, he was "victim of his own success" and of his skills, since, in 1657, the government of the city, after indications from the Pope himself, appointed him *Superintendent of Public Waters*, with the important assignment of directing the monitoring and maintenance of the flow of rivers and the stability of the bridges and of the canals. The prestige of this office derived from the importance of the watercourses as the main and fastest commercial routes of the time, but with this also came, as a poisoned apple, the "the problem of the water".

The so-called "problem of the water" was a very delicate issue involving the towns of Ferrara, Ravenna, and Bologna and the Po, the primary and longest Italian river. Cassini himself explains this problem in his memoirs. Stellata was a village near Ferrara, which today pinpoints the junction of the administrative borders of the three regions of Lombardy, Veneto and Emilia-Romagna. Here the river Po splits in two branches. One of them received the waters of two tributaries, the Panaro and the Reno (Fig. 8.2), and continued towards the city of Ferrara, where it split in two more branches, called *Po di Volana* and *Po di Primaro*. Nowadays, the river Reno is not a tributary of the Po anymore, but at that time it represented the access of the city of Bologna to the Po and to all the trade with the many other cities connected by this link. The *Po di Volana* and *Po di Primaro* eventually reached the Adriatic Sea, but first a canal toward another city, Ravenna, departed from the latter branch (Fig. 8.3).

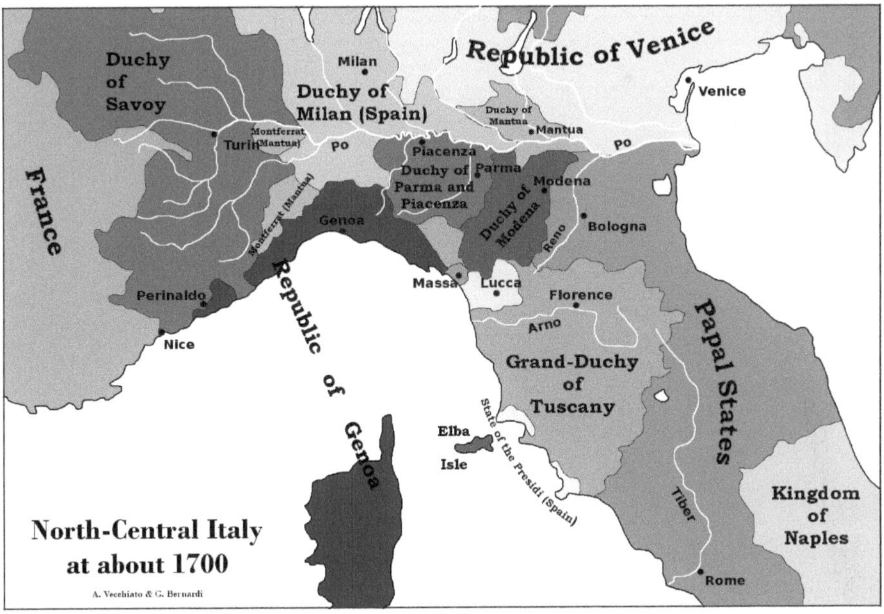

Fig. 8.2 North-central Italy around 1700

This configuration put Ferrara at the crossroad of many important commercial routes. One was from Bologna, but the most important ones were those arriving at Ferrara from Lombardy. This convenient configuration, however, had been put at risk by frequent floods, which caused continuous changes in the course of the river and interfered with the artificial watercourses. In particular, this was because in the previous century these modifications had started draining the Primaro branch and therefore the canal to Ravenna. Moreover, at the Stellata branches, the Po started to convey its waters into a large riverbed called *Po Grande*, which went directly to the sea north of Ferrara.

Many attempts had been undertaken to regulate the watercourses and preserve the commercial routes, but they were failing in the goal of satisfying the economic interests of all the involved cities. In practice, the already considerable technical difficulties had been made even worse by their resultant commercial consequences, making this more a diplomatic, rather than an engineering, problem. For example, some projects, like one of the architect Aleotti or, in a different way, another one proposed by engineers from Ferrara, would have interrupted the access of the Reno River to the Po, thus penalizing Bologna.

In May 1657, Cassini initiated the first inspection in the territory of Ferrara, accompanied by the Marquis Tanara, ambassador of Pope Alexander VII. This was only the beginning of a complex history composed of measurements, reports, travels, and conferences, which lasted for many years as evidenced by the large number of letters and documents that survive to the present day. In 1662, for

Fig. 8.3 Map of the Po River close to Ferrara in a map from 1568

example, the Marquis Cornelio Malvasia published "*Ephemerides Novissimae Motuum Coelestium*", namely his ephemeris calculated with the Cassini tables obtained from the heliometer data of San Petronio. This was a very important work at the time, which also included a solemn poem in Latin by Cassini (a practice inherited from his past studies with the Jesuits) and four letters on astronomical subjects between Malvasia and Cassini. At the end of the second letter, Cassini informs the Marquis about his next "water" travel to Ravenna and Ferrara, where the astronomer complains that this burdensome task will take him away from Bologna for quite a long period.

Despite the heavy burden due to this assignment, the Bolognese scientist did his best to keep working on astronomical problems. During a stay in Ferrara, Cassini observed a solar eclipse, and Cassini used a method invented by him to represent, on a map, the different appearances for all places on Earth. As he reports in his memoirs, this method had been devised earlier, during another eclipse, in the presence of Francis, Duke of Modena, but the inquisitor of city was concerned by this innovation. Unfortunately, it is not possible to have a better understanding of what exactly alarmed the inquisitor in the writings of the scientist, but at that time Cassini was not allowed to publish his technique.

In addition to this eclipse, in Ferrara, Cassini also accurately observed the star that represents the "shoulder" of the Auriga constellation passing at the zenith, and his data were inserted in the work of Father Riccioli, entitled "*Géographie réformée*".

In June 1658, Cassini presented his plan at the Congress of Ravenna, a board of religious and civil authorities that had the task of examining the technical proposal and to make executive resolutions. The board's decisions, obviously, were influenced by other factors like the economic and political aspects of the proposals. Cassini, as part of his assignment, knew it very well, and his design tried to regulate the Reno River without hindering Bologna or Ferrara. Despite his hard work, however, this proposal was declined, mainly through the hesitation of the Pope, who had to make the ultimate decision and was not able to decide himself. The discussion dragged on for a couple of years, at the mercy of ambiguous attitudes and opposing interests, which embittered the young scientist, but he could at least undertake some initial actions, like the levelling of the sluice of Casalecchio, a small city close to Bologna. Here Cassini used the measurements to estimate also the length of a degree along Earth's meridian, which is eventually connected with the size of our planet, but unfortunately the tools were inadequate, and the work would be accomplished many years later, when the appropriate equipment became available.

Because of his duties as Superintendent of the Public Waters, Cassini had to make several trips to Rome to report Pope Alexander VII, which had both positive and negative effects for his career. Indeed, this privileged link increased his fame and estimation in the academic world, in the clerical entourage, and among the nobility, especially with the Chigi, the family of origin of the then current Pope. On the other hand, these contacts made him even more exposed to non-astronomical tasks. In this period, not only Alexander VII personally submitted to Cassini various types of assignments and questions, but also his brother, the Prince and general Mario Chigi, gave him the responsibility of overseeing the renovation and the extension of the Urbano Fort, a fortification between the Papal States and the Duchy of Modena. Here, Cassini discovered a water source that he elevated to reach the embankment with an ingenious system of pipes, and, on August 18th of 1663, he observed a lunar eclipse, which indicates how he always strived to pursue his astronomical research.

In 1664, he began a new effort concerning the waters. The Pope asked Cassini to govern the course of the river Chiana, a little and unstable waterway, which sometimes was a tributary of the Tiber, the river of Rome, and sometimes of the Arno, the one of Florence. Once again, despite two years of measurements and reports, the solution presented by Cassini was not pursued, but this assignment had also a positive implication. Vincenzo Viviani, mathematician and pupil of Galileo Galilei, represented the Government of Florence, and this duty nurtured a mutual esteem and a friendly relationship between the two scientists that lasted until the death of Viviani.

Curiosity

The continuous travels for field surveys in Ferrara and Ravenna, and for his reports in Rome, prevented Cassini from fulfilling his regular teaching duties in Bologna; in the State Archives there is a document that testifies to the exemption of the astronomer from teaching because of an assignment of public character. Without this resolution by the University board, Cassini would have been constantly fined for having missed his lessons without just cause. At that time, in fact, a public official called "*ufficiale puntatore*" was appointed specifically to guarantee the regular development of the lessons, and he received a fee for each academic irregularity that he reported.

During the river-Chiana assignment, Cassini also made observations as biologist and naturalist, writing important considerations about the lifecycle of a species of gallflies, a family of insects also called "Gall wasps" or "Oak galls", which induce the formation of galls on oaks to sustain the development of their larvae. In his notes, there are also reports about a deposit of shells close to the mountains that he supposed to be of ancient origin.

At the time, it was thought that the Scaffaiolo Lake was so deep as to make a channel to Bologna, and Cassini applied a rigorous investigatory method to verify this.

Where was Cassini staying when in Bologna? As reported in his memoirs, in the intervals between his journeys as Superintendent of Public Waters, he resided with the Marquis Angelelli, who prepared a room with scientifically- and mathematically-inspired paintings and other decorations.

Chapter 9
The Queen and the Comet

...her letters were not just chit-chat

the Minerva of the North

"*She was a woman with a fiery nature and of many contrasts, ranging from violence to piety, from good to bad, from superstition to the science. Nonetheless, her defects were greatly exceeded by her many virtues which earned her the title of the Minerva of the North and obtained for her a permanent place in the history of the cultural life of Rome.*" This is the description of Christina of Sweden given by the American historian Silvio Bedini in "*Making Instruments Count. Essays on Historical Scientific Instruments Presented to Gerard l'Estrange Turner.*"

We have already met this remarkable woman in the chapter about the meridian line, or heliometer, of S. Petronio in Bologna. Her position, at the time, was at the center of a very important political and religious affair. Christina, in fact, had ascended the throne of Sweden in 1632 at the age of six. Because of her young age, she initially reigned under the guidance of the state chancellor, gradually becoming more and more independent until taking full control in 1650. In 1654, however, she announced her abdication in favor of her cousin Charles X Gustav. This is not the place for a detailed discussion of her life, but at the root of this resolution were several reasons. Among them were the physical and psychological stress after more than 20 years of reign, and the refusal of a marriage after which she would have ceded her role to the husband, but also the hostility of the aristocracy, whose power she had tried to limit, and finally her conversion to Roman Catholicism in a Protestant nation (Fig. 9.1).

Granted a substantial annuity, she then moved to Rome, also stopping in Bologna where, as previously mentioned, she met Cassini who gave her the first drawing of the meridian line of S. Petronio, among the other gifts and attentions she received in the city. When she finally arrived in Rome, as Bedini reports: "*Inside the gate she was welcomed by the entire College of Cardinals mounted on their mules. Farnese offered Christina a saddled horse on which she rode with her favorite black scarf and plumed hat, her appearance contrasted sharply with that of Roman nobility with their velvets and diamonds. Surrounded by cardinals and followed by a train of glittering carriages, the cavalcade wended its way to the*

G. Bernardi, *Giovanni Domenico Cassini*, Springer Biographies,
DOI 10.1007/978-3-319-63468-5_9

Fig. 9.1 Christina of Sweden (1626–1689) oil painting on canvas, 1661 (Open Image Archive of The Royal Armory, Stockholm)

Basilica of St. Peter. The thunder of cannon from Castel Sant'Angelo and the cheers of the clamoring populace heralded its progress as it advanced slowly through the city toward the Vatican".

Her journey through Italy to Rome had been planned in detail by the Vatican (her conversion was a political affair); and the Queen changed her name from Kristina in Christina Alexandra, the last chosen in honor of the reigning pope, Alexander VII, even if the name also commemorated one of her heroes, Alexander the Great.

The official entry into Rome took place in 1655 in a couch designed by the famous architect Gian Lorenzo Bernini, who became a lifelong friend of hers. Such friendship was built on a solid basis: The queen was one of the best educated women of the century and "*She was exceedingly learned and keenly interested in theology, philosophy, and alchemy*" as Alan Cook recalls in his "*Ladies in the Scientific Revolution*". She was also called the *Minerva of the North* because Christina's passion for paintings, manuscripts, sculptures, and books had attracted many philosophers in Stockholm. In a similar fashion in Rome, she became a leader of the theatrical and musical life, protected many artists and composers, and encouraged the development of science, as a founder of the Royal Academy in 1674, a science society at the Riario Palace, her abode, along the river Tiber.

Une Comète Proche du bec du Corbeau

As Cassini writes in his memoirs, during the stay in Rome, at the end of 1664, there appeared "*... une comète proche du bec du corbeau*" that is, a comet next to the beak of the Crow constellation (another one became visible in April 1665), and, contemporaneously, Christine of Sweden reappeared in the life of the Italian astronomer. Indeed, the former queen remembered very well the young astronomer she had met in 1655 in Bologna and his particular gift. Citing again Bedini's work: "*Christina's interest in Cassini, which was to span her lifetime, is not surprising. A capable mathematician and unquestionably the foremost astronomer of his time and extremely successful in his work, he was apt to create situations to take the best advantage of them. He had a simple manner, was restrained in his speech, and timid even when he knew better than others, always affable, and a master of the art of complimenting*" (Fig. 9.2).

Prince Mario Chigi, the Pope's brother, notified Cassini of the appearance of the comet, so he could observe it regularly beginning on December 18, 1664, from Chigi Palace, aided by an abbot to record its position with respect to the nearby stars. The cometary path was then represented on the prints by adding small grains of lead to the print master of a stellar map. This comet and the next one, in 1665, are scientifically very important because their regular observation and the consequent calculation of the ephemerides allowed Cassini to formulate his theory about cometary motion, started years before in the Panzano observatory but never finished. His patron, the Marquis Malvasia, died in March 1664, but their research continued with the observation of the newly-visible comets with the help of new collaborators.

In Rome, this observational program continued not only from the Chigi Palace, but also with the Queen, in her residence. At her invitation, the astronomer and Cardinal Assolini (who would become her sole heir) went here every day to pay a visit to the noble woman, and, before the observations, they discussed science to the great pleasure of Christina. In this regard, Cassini reports in his memoirs that, in these discussions, the Queen was often inclined to side with his opinions so, as man of the world, the Italian scientist was used to try to move the conversation to topics that would be more pleasing to the eminence.

Fig. 9.2 Depiction of the comet of 1664 (Max Planck Institute for the History of Science, Berlin)

His attempts, however, had little success, since most of the time the Queen was quite curious about the comet, and she wanted to follow it with Cassini. She made clever comments and posed questions, for example, noticing its high speed to the northwest, supposing that the comet would complete an entire revolution of the sky in a short time. But Cassini, in his memoirs, remembers having to explain to her that the comet movement had to appear to slow down and become stationary.

She was very surprised, and after further requests for explanations, he exposed in more detail his work and asked her permission to dedicate it to her. This work was published in May 1665, in Rome, with the Latin title "*Theoria motus cometae anni MDCLXIV*". It collected the observational data of both the 1664 and the 1665 comets—the former from 18 December to 15 January, the latter dating to April of that year—and Cassini supposed that the latter was the same described by the Danish astronomer Tycho Brahe in 1577. As promised, his dedication reads "*Christinae Alexandrae Suecorum Reginae Augustae*", the Queen Christine of Sweden, and also ends with: "*Fulgeat interim diutissime Romano Coelo Maiestatis*

Tuae Sydus", namely that the Queen star would long continue to shine in the Roman sky.

This work is very important because it is the first that reports a precise sequence of observations and the entire path of a comet, also supported by mathematical calculations. Cassini, indeed for the first time in the history, tried to determine with the best possible accuracy the cometary movement, and, in this attempt, he employed the same laws used for the planets. The importance of this attempt should not be underestimated. It has to be remembered, in fact, that, according to the Aristotelean and Ptolemaic models, the comets did not belong to the heavens, but they rather had a terrestrial origin. Tycho had already questioned this interpretation, precisely with the previously mentioned comet of 1577, but the transition between the classical, geocentric, and Aristotelean physics and astronomy and the new heliocentric model, was still an ongoing process, and had to await Newton for its completion.

The involvement of Cassini in such philosophical debate was not "political", but this and all his other behavior on the scientific side indicates his adherence to the new scientific models. Once again, as happened with the meridian line, his cautious conduct was probably the result of both his dependency within a potentially hostile environment and of his will to stick strictly to scientifically viable statements. How much this demeanor was appropriate can be realized by the reactions to this work; the difficulty to accomplish this task, and probably also the resentment of other scientists made it difficult to be widely accepted and exposed Cassini himself to some criticism.

Curiosity

The Pope was aware of Cassini's work about comets and asked him for the dedication, but Cassini promptly informed him that he had already promised it to Queen Christina, thus receiving the approval of the Catholic sovereign.

Cassini reported in his memoirs that his abode in Rome, during the period as superintendent, was the Colonna Palace, with the Marquis Campeggi, Ambassador of Bologna. Here, usually after dinner, a carriage with a page boy brought him to the Queen at Riario Palace where they talked of science, waiting for the night to observe the comet. Cassini writes also in his memories that in the presence of the Queen, he was bareheaded, so she, "*avait la bonté*" (had the goodness) to protect him from the cold night air with a scarf.

Riario Palace, located in Via della Lungara in Rome, became Corsini Palace and today is the headquarters of the *Accademia dei Lincei*, or Lincean Academy, founded in 1603. Literally, the meaning of its name is the Academy of the Lynx-Eyed. It was one of the first scientific academies in Italy and was named after the lynx, an animal whose sharp vision symbolizes the observational prowess that science requires.

Chapter 10
And the Winner Is…

And thence we came forth again to see the stars.
Dante Alighieri, Inferno XXXIV, 139

"*S.M. ayant fait travailler en vain à un miroir concave de verre, j'en fis venir un très-grand que j'avais chez moi à Bologne, et je le laissai entre les mains de la Reine* [Christine of Sweden]. *Je ne sais ce qu'il est devenu depuis. [...] Je l'ai toujours regretté, ne croyant pas qu'il y en ait eu un, ni plus grand, ni meilleur, de cette matière...*" [S.M. had in vain made to work a concave mirror of glass. Finally, I managed to bring here a large one I had at home in Bologna, leaving it in the hands of the Queen [Christine of Sweden]. I do not know what has happened since then. […] I always regretted it, not believing that there has been one larger or better made of this material].

These lines testify two lessons learned at a high price by Cassini: first, never give anything if it is so unique; second, valid optical artisans have always been a valuable resource for astronomers. Indeed, these figures would play a fundamental role in the scientific achievements of the Italian scientist. The high precision observations of the comet of the years 1664 and 1665 would have not been possible without good instrumentation, and these years would be critical for Cassini not only the regarding comets but also Jupiter, as we will see in the next chapter.

In 1610, Galileo discovered with his telescope four Jupiter satellites, which he called *Medicean stars*. The great Pisan, sensing the great importance of an accurate determination of their motions for the problem of the longitudes (a tricky problem related to the navigation that will be solved the following century by Harrison in a completely different way) entrusted his favorite pupil, Vincenzo Renieri, with the task of continuing his work on this subject. This monk, however, died suddenly in Genova in 1647 without having been able to produce the tables with the ephemerides of the Jovian satellites.

Probably during his studies in Genova, Cassini, who mentions the name of Renieri in his biography, could see the sketches of the unfinished tables, but he also believed that, in order to finish this very important work, he needed time and, above all, suitable equipment.

G. Bernardi, *Giovanni Domenico Cassini*, Springer Biographies,
DOI 10.1007/978-3-319-63468-5_10

Campani versus Divini

At the time of Cassini, the best manufacturer of telescopes of a European level was Eustachio Divini (1610–1685), but the younger Giuseppe Campani (1635–1715) was matching, if not exceeding, him (Fig. 10.1). He was able to make high quality optical lenses, probably because of the introduction of new techniques, which however he kept secret forever. This rivalry became an actual challenge, which occurred over the years 1663 and 1664. Such kinds of competition were typical of the Florentine scientific environment, and were called "paragoni", an Italian word meaning "comparisons". In fact, they consisted of direct comparisons between tools under the same "experimental" conditions, and, in the presence of princes, patrons, and a few specialists, among whom in this case was Cassini.

As Maria Luisa Righini Bonelli and Albert Van Helden reported in "*Divini and Campani: a forgotten chapter in the History of the Accademia del Cimento*" (Annali dell'Istituto e Museo di Storia della Scienza, Firenze, 1981) the first competition took place in Florence in October 1663: "*This first open "paragone" [competition] was held at the residence of Matthias de' Medici, and it was attended by a large number of virtuosi. High officials of Church and State mingled with scientists, among whom Giovanni Domenico Cassini, professor of astronomy at the University of Bologna, was the most distinguished. The results of the comparison were mixed: it was judged that Campani's telescope showed objects more clearly, while Divini's instrument magnified more*".

Another competition followed on April 30, 1664, organized in Rome in the presence of Cassini, who never openly disapproved the lenses of Divini, but, on several occasions, he demonstrated a preference for those of Campani. For the sake of clarity, however, and to help the reader understand the highly competitive mood of the participants, it should be pointed out that this comparison: "*had, again, been carefully arranged by Matteo* [Giuseppe Campani's brother] *and Giuseppe* [Campani] *to put Divini at as much of a disadvantage as possible*".

The above statement can be better explained by stressing that in these disputes the instruments played only part of the game. A very important role was also that of the people involved in the competition. For example, Divini was too submissive, contrary to the brothers Campani, who always played as a team, with Giuseppe flanked by Matteo as his spokesman. They also constantly tried to create more favorable conditions for their tools; for example, before the experiment in Rome, they firmly mounted their telescope on its support, while that of Divini was resting precariously balanced on some chairs.

So, at the end of 1664 "*...the war was over, and Campani was the winner. His reputation continued to grow as announcements of celestial discoveries made with his telescopes came from Italy in a steady stream...*" as reported Righini Bonelli and Van Helden. The collaboration between Cassini and Campani, the astronomer and the artisan of telescopes, would go on for a lifetime, ennobling them both: "*Cassini's many discoveries, not Prince Leopold's impartial tests, gave Giuseppe Campani his place in history*".

However, it would be too simplistic to define Giuseppe Campani as just a skilled builder of telescopes. Actually, he was also an observer, as evidenced by the various letters between them, in which they exchanged their observational results

Fig. 10.1 Portrait of Eustachio Divini [Wellcome Library, London. Wellcome Images (CC-BY 4.0)]

about the satellites of Jupiter or the rings of Saturn. Nonetheless, their roles were well defined, as once again Righini Bonelli and Van Helden wrote: "*The discoveries came in two great waves, the first from 1609 to 1612, and the second from 1655 to 1685. In the first Galileo was the dominant figure because the combination of his telescopes, his eyes, and his mind was the best by far. In the second period, Cassini and Campani were dominant figures because the combination of Campani's telescopes and Cassini's eyes and mind was the best observing system*".

In this period, then, Cassini was still busy with the public waters affair, but finally he managed to find a valid collaborator in Campani. His instrumentation would become fundamental to his systematic study of the planets, and it enabled, for example, the observation of the shadows of the satellites of Jupiter on the disk of that planet. Using the exact words of the astronomer: "*Au milieu des occupations que me donnaient les affaires publiques, je faisais la nuit des observaions astronomiqué avec une excellente lunette que m'avait donnée M. Campani, qui avait communiqué au public la découverte que j'avais faite des ombres des satellites de Jupiter sur le disque de cette planète; ce qui avait engagé d'autres astronomes à les observer*". [While I was kept busy by the assignments of public character, I could make astronomical observations at night with an excellent telescope, which had been given to me by M. Campani, who had communicated to the public the discovery I had made of the shadows of the satellites of Jupiter onto the disc of this planet, which I had engaged other astronomers to observe.] But, as reported in the last sentence, many astronomers would then also observe them, and some were not too benevolent, as we will see shortly.

Curiosity

It was not uncommon to see turrets or terraces for astronomical use rising up from the buildings of the noble and powerful Roman families. This is because these families surrounded themselves with scholars to promote their scientific research, and they even participated to the astronomical observations.

On August 6, 1664, Cassini observed a lunar eclipse in Rome from the *Propaganda Fede* (Faith propaganda) College with a telescope of Giuseppe Campani. His observational data are archived in Florence, in the National Library (Galileiano 272, c.33).

Giuseppe Campani was not only an optician but also an astronomer and his astronomical observations and the descriptions of his instruments are recorded in his papers: "*Ragguaglio di due nuovi osservazioni, una celeste in ordine alla stella di Saturno, e terrestre l'altra in ordine agl' instrumenti*" (Rome, 1664, and again in 1665); "*Lettere di G.C. al sig. Giovanni Domenico Cassini intorno alle ombre delle stelle Medicee nel volto di Giove, ed altri nuovi fenomeni celesti scoperti co' suoi occhiali*" (Rome, 1666).

Several decades after his death, in 1747, the workshop of Campani was donated to the Academy of Science of Bologna.

Chapter 11
Jupiter et al.

> *...bellissima, ed atta a convincere molte opposizioni*
> [...beautiful, and ready to convince any opposition]
> *Prince Leopold letter to Michel Angelo Ricci,*
> *Florence 26 August 1665*

In 1665, Cassini was frequently traveling in Tuscany or Lazio, still committed to assignments of public interest on behalf of the Pope. However, on each travel he brought with him his new Campani telescope and managed to observe with this instrument during the night. As anticipated in the previous chapter, this activity led to the detection of the shadows that the satellites of Jupiter were projecting onto the disk of the planet. These satellites had been discovered by Galileo in 1610, who named them the Medicean planets after the family name of the grand duke of Tuscany. They also had mythological denominations, which eventually prevailed: Io, Europa, Ganymede, and Callisto, namely the lovers of the god Jupiter in Greek mythology. Soon afterwards, Campani announced the discovery of these shadows, so several astronomers began to observe Jupiter in search of them (Fig. 11.1).

A controversy about the results of these observations started quite soon. First, someone from Rome observed one dark shadow and another less dark one. Cassini clarified that the latter was not the shadow of a satellite but rather a physical spot on the planet. To prove his statement, as reported in Cassini memories, he observed that it did not follow the movement of any orbiting body, and that it appeared every nine hours and 56 min, which the astronomer attributed to the rotation period of the planet. This triggered an intense debate about the paternity of this discovery.

For example, Father Gottignez, mathematician of the Jesuit College in Rome, claimed to have done the first observations of this phenomenon, and Cassini answered with a printed letter on August 8, 1665. In support of the statement of the Italian scientist, Abbot Ottavio Falconieri published the letters between Cassini and himself as proof of his past observations, and Cassini made public his predictions on the return of the spot on Jupiter, to dispel any doubts.

From Tuscany, the mathematician of the court, Giovanni Alfonso Borelli, confirmed the above predictions. Prince Leopold, Grand Duke of Tuscany, then

G. Bernardi, *Giovanni Domenico Cassini*, Springer Biographies,
DOI 10.1007/978-3-319-63468-5_11

wrote to another mathematician, Michelangelo Ricci, a former student of Galilei and friend of Torricelli, defining Cassini's predictions as: "*...beautiful, and ready to convince any opposition*". His comments had probably been done to contradict the words of criticism that had started to circulate about this discovery. Some competitors, in fact, had considered Cassini's observations of poor accuracy, on the basis that they had been conducted during his travels, rather than from a comfortable observatory.

On the contrary, the famous Dutch scientist Christiaan Huygens wrote a letter of appreciation to Prince Leopold in Latin: "*Cassini autem novam observationem quod attinent de Jovialium umbris, ea plane egregia ac felix mihi visa est, neque de rei veritate dubitandum putavi, quemadmodum ab aliis fieri intelligo; ac minus etiam, postquam ipse die 26 septembris anno praeterito 1665 umbram Comitis III quam Cassinus, comparituram praedixerat, manifesto observassem...Parisijs, 22 Junij 1666*" from Angelo Fabroni "*Lettere inedite di uomini illustri*", *Firenze 1773*. In English, "But Cassini [made] a new observation of the Jupiter shadow, which certainly seemed to me admirable and very successful, and I never doubted of its truth, as I hear from others; and even less after, on September 26, 1665, as predicted by Cassini, I clearly observed the shadow of the third companion (satellite, A/N)... Paris, June 22, 1666".

Despite the inconveniencies of his travels and of the scientific controversies, the new equipment by Campani enabled Cassini to resume his astronomical work with renewed energy, devoting himself to the planets. In particular, as already mentioned, his main targets of this period were Jupiter and its satellites, with special attention devoted to the rotational motion of the giant planet and its spots. Indeed, he published in 1665 the table of the "revolutions" of Jupiter spots: "*Tabulae quotidianae revolutionis macularum iovis*" (archived in the Astronomy Library of Bologna) after the "Astronomical letters Falconieri". He was also the first to observe and describe the atmosphere of the planet, noting its changeability over time. These studies will be followed centuries later with more powerful tools, but probably he was the first, or at least the second, who observed the most prominent and well-known characteristic of the largest planet of the solar system, namely its Great Red Spot. The uncertainty comes from the fact that Robert Hooke, a British philosopher, described a spot on Jupiter in May 1664, but his report places it in the wrong belt. On the other hand, the description of Cassini one year later of a "permanent spot" is much more accurate and unambiguous. Ultimately, it is not possible to state clearly the precedence of Hooke over Cassini in this respect.

Apparently, the study of Medicean planets, namely the satellites of Jupiter, was proceeding with some difficulties because of the commitment of the Italian scientist as Superintendent of the Public Waters, as evidenced a letter of January 9, 1666, sent by Cassini to his colleague Viviani. Previously, Viviani had sent the work of Galileo to Cassini, so the latter could make a comparison with his own measurements. However, as we can read in the letter, the official appointments left him little time for his astronomical research. Luckily, he could use every second of this little time since he was still young and in good health. In addition, a natural talent as an

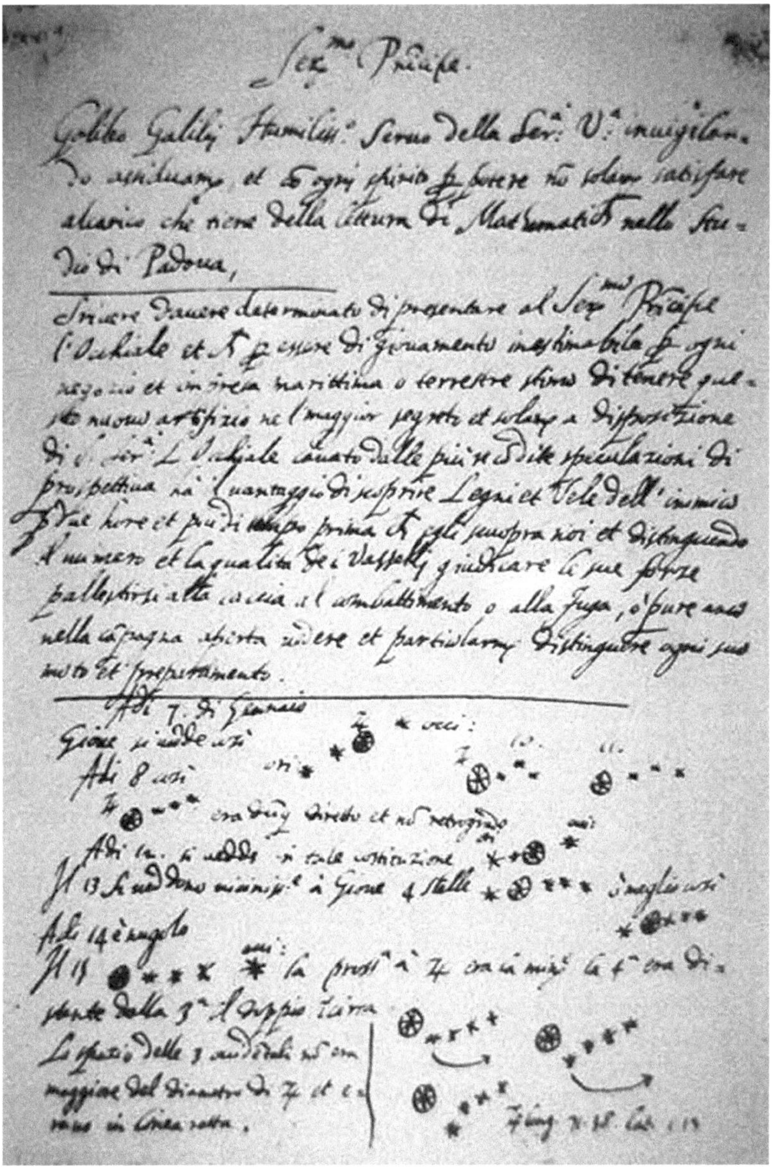

Fig. 11.1 A draft letter written by Galileo Galilei in August 1609 to Leonardo Donato, Doge of Venice. In the lower part, he noted (in January, 1610) his first observations of the planet Jupiter and four of Jupiter's moons

observer and his ability to perceive the essence of celestial phenomena helped him to make the most of such research work.

Jupiter, however, was not the only planet observed by Cassini. Already in 1666, Mars captured his interest as well, and it was observed with a Campani telescope, with particular attention to the "spots" visible on the planetary surface. The interest in these features had the same motivation that had the Great Red Spot on Jupiter: they could be used to determine the spin of the planet. Indeed, his first publication on the subject: "*Martis circa proprium axem revolubilis observationes Bononiae habitae*" [Observations of Mars revolving about its own axis made in Bologna] was exactly about the determination of the spin of Mars. It contains original drawings by Cassini showing the different positions of various spots observed on Mars in February, March, and April of 1666. Simply following their movement from east to west, observing their disappearance and then their reappearance, Cassini was able to estimate a period of rotation of the red planet of 24 h and 40 min, just three minutes longer than the presently accepted value.

From Rome, Campani was using his most powerful telescope, with a focal length of 50 "*palmi*" or palm (an old unit of measurement) corresponding to about 10 m and confirmed the observations of Cassini. This result would be contested by two Roman scientists who, using a Divini telescope, estimated a rotation period of 12 h. The *Journal des Sçavans* in Paris, the earliest academic journal in Europe which later was renamed *Journal des savants*, and the *Philosophical Transactions* in London, reported the controversy, and Cassini, in order to defend the accuracy of his calculations, published the "*Dissertationes astronomicae apologetticae*" (Fig. 11.2).

His detractors, in turn, released "*Martis revolubilis observariones romanae ab affectis erroribus vindicatae*" or "Roman observations of the rotation of Mars redeemed from the mistakes from which were encumbered", but Cassini's work proved to be exact and the present day accepted rotation period of Mars is 23 h 37 min and 23 s.

Curiosity

The "*Accademia del Cimento*" (Academy of Experiment) in Florence was the first society that explicitly and systematically used experiments as a method of scientific investigation. It was founded by some students of Galileo Galilei, but received considerable support from the government of Tuscany, and, in particular, from Prince Leopold. The latter used to send Cassini many problems about the equilibrium of liquids, which the scientist always resolved. In his memoirs, he recounted that the prince, in recognition of these brilliant results, gave him the first place at his side on the occasion of a visit of the astronomer to Florence. Cassini, also knowing his interest in astronomy, gave the Prince a model of his planetary system.

One of the many interests of Cassini was medicine, so one should not be surprised to find that he worked on blood transfusions, a hot topic in Italy and abroad. His name appears in some letters that cited successful experiments on lambs in Bologna, on May 28, 1667, in the house of Cassini, where he received students and colleagues.

LE

IOVRNAL

DES

SÇAVANS

Du Lundy V. Ianvier M. DC. LXV.

Par le Sieur DE HEDOVVILLE.

A PARIS,

Chez IEAN CVSSON, ruë S. Iacques, à l'Image de S. Iean Baptiſte.

M. DC. LXV.

AVEC PRIVILEGE DV ROY.

Fig. 11.2 Title page of Journal des sçavans of Monday, 5 January 1665

In the Billings microscope collection at Walter Reed Army Medical Center in Washington, there is a compound monocular microscope on a tripod made by Giuseppe Campani.

Pope Alexander VII was another important personality who was very interested in astronomy, and Cassini had frequent talks with Cassini. Alexander confided to him that, when he was young, he had enjoyed building portable sundials, and Cassini gave him a world map as a gift.

Cassini was not the only one to observe the spots on Jupiter with the most powerful instrumentation: Giuseppe Campani disputed with his competitor, Eustachio Divini, the priority of this discovery.

As previously mentioned, in Rome Cassini used to spend time with the most powerful families of the city, and therefore of the Papal States, like the Chigi, Barberini, and Colonna. A common attendee during his observations was Maria Mancini Colonna, the most famous of the nieces of Cardinal Giulio Mazzarino, Chief Minister of France. She lived with her sisters at the French court, and Maria was lively and beautiful. The young King Louis XIV became infatuated with her and wanted to marry her, but her uncle gave Maria in marriage to Prince Lorenzo Colonna. It seems that she was devoted to astronomy, and Cassini dedicated to her some verses in Italian describing the constellations, which she enjoyed memorizing. These astronomical verses and others are preserved at the Library of the Paris Observatory, and you can find some examples of them in Appendix A.

Chapter 12
Ephemerides Bononienses Mediceorum Syderum

... l'observation d'un phénomène extraordinaire...
[the observation of an extraordinary phenomenon]
Giovanni Domenico Cassini

On June, 20, 1667, Cardinal Giulio Rospigliosi, whom Cassini had already known during his frequentation of the court of Alexander VII, ascended to the Papal throne, taking the name of Clemente IX.

The next year, in March 1668, a "*spina celeste meteora*", as it was defined by Cassini who did not know what it was (literally, a "heavenly meteor spur"), appeared in the sky, moving through the constellation of Eridanus towards that of Orion. According to what he wrote in his memoirs "*... au même endroit du ciel et avec le même mouvement qu'un phénomène tout semblable qui avait été observé du tems d'Aristote, et qui a paru de nouveau au même lieu, l'an 1702.*" That is, a very similar phenomenon had been observed back in the days of Aristotle in the same place in the sky and with the same motion. The same was to appear again in the same place in 1702.

Various astronomers, in various places in Italy, saw a trail of light on the horizon, but, after nine days, Cassini could not tell if it was a new comet. It wasn't. Actually he was observing a phenomenon known as *zodiacal light*, namely a faint glow barely visible on moonless nights, roughly in the ecliptic region (the zodiac) produced by the light of the Sun scattered by dust particles (Fig. 12.1). The Italian astronomer was one of the first to investigate this event, and he published his observations in the same year. Eventually, some years later, he would coin himself the name we still use to this day, and, according to some sources, he also grasped the currently accepted explanation in terms of dust-scattered light.

In the same year, Cassini circulated an astronomical work that would become very important not only from the astronomical point of view, but especially for the life of the Italian scientist. Finally, he managed to publish the ephemeris of the Medicean satellites with respect to the meridian of Bologna: *Ephemerides Bononienses Mediceorum Syderum.* Printed in Bologna, the book was dedicated to Cardinal Giacomo Rospigliosi, nephew of the new Pope. The importance of this

G. Bernardi, *Giovanni Domenico Cassini*, Springer Biographies,
DOI 10.1007/978-3-319-63468-5_12

Fig. 12.1 Artistic representation of the zodiacal light from the book "Astronomy for the Use of Schools and Academies" by J.A. Gillet and W.J. Rolfe (New York: Potter, Ainsworth, & Co., 1882)

work about Jupiter and its satellites was already envisioned by Galileo about 50 years before, because it represented a possible way to solve the so-called "problem of the longitudes" (Fig. 12.2).

In short, while the determination of the latitude of a site on the Earth was a quite simple problem, the estimation of the longitude was much more difficult. At the

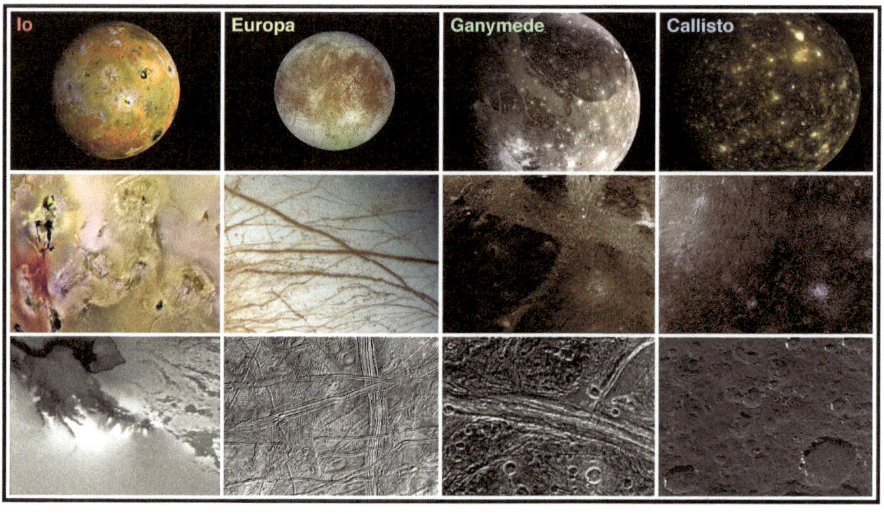

Fig. 12.2 Global and high-resolution images of Jupiter's moons from the Galileo and Voyager spacecrafts (NASA/JPL/DLR)

same time, it was essential to draw precise maps of the Earth, and therefore for navigation, commerce, and war.

The great Pisan had understood that the *solar* times of the ephemerides of the Medicean satellites had to be different when observed from different locations on Earth. This happens for simple geometrical reasons. The solar time, in fact, is the time associated to the position of the Sun in the sky, and thus it depends on the longitude of the place. This can be easily understood, e.g., by realizing that noon is *defined* as the time when the Sun crosses the local meridian. Therefore, because of the spin of the Earth, when in a certain location it is noon, in an eastern location it will be afternoon, while in a western one it will still be the morning.

This is the same concept at the basis of the time zones we use today: the local time depends on the longitude, and thus a difference in local time between two places can be converted into a difference in their longitudes. Since the Earth makes a complete turn around its axis (360°) every 24 h, this conversion factor is approximately 15° per hour. Obviously, the same reasoning can be applied not only to the Sun, but also to any astronomical event, such as the ephemerides of the Jupiter satellites. For example, if a satellite occultation (i.e., the satellite transit behind the edge of Jupiter) were observed at midnight from Bologna, the solar time of the same occultation would be about 36 min earlier in Paris because the longitudinal difference of the two cities is approximately 9°. In practice, the idea of Galileo was to use the occurrence of specific events of the Jupiter system, like the frequent occultations of its satellites, as a clock.

Once their ephemerides were known for a specific location on the Earth and made available on suitable tables, then one could observe the satellites from another place, determining the longitudinal difference with that of the reference site simply by recording the local solar time of specific events and comparing them to that in the tables.

As recalled earlier, Galileo had entrusted Vincenzo Renieri, one of his favorite pupils, with the task of preparing such tables, and later, after his premature death in 1647, Cassini took this task upon himself. However, Galileo, Renieri, and Cassini were not the only ones who tried to use this method for the estimation of the longitude. Several mathematicians in Europe were also dedicated, but no one was able to complete the needed tables. It was Cassini who, after years of observations and calculations, would finally determine the daily motions of the satellites of Jupiter, and the times of the eclipses (occultations) referred to by the meridian of Bologna.

On July 28, 1668, the *Giornale dei letterati di Roma* (Journal of the Scholars of Rome), announced the publication of the "*Ephemerides*". Cassini's tables spread quickly throughout Europe since, as noted previously, they had potentially huge practical utility. It was possible to determine the geographic coordinates of any point on Earth, that is, calculate the longitudes depending on the timing of the eclipses of Jupiter satellites, and to correct the charts in use to obtain more reliable maps. This was very important and strategic for navigation, although the application of this method at sea, during navigation, was still too difficult. The solution to this last problem had to await Harrison and his nautical chronometers, in the next century.

The activities of Cassini were constantly and carefully monitored from France, by none other than Jean-Baptiste Colbert, the most powerful man in the kingdom after Louis XIV. Therefore, when the work on the ephemerides became a fundamental text for the drawing of geographical maps, Colbert, as a first step, invited him to join the *Académie Royale des Science* (French Academy of Science). In return, Cassini periodically sent his works to France.

Such kind of contacts with prominent European scientists were not unusual. Indeed, at the time, France was involved in a large program of hiring top scientists. For example, Christiaan Huygens, prominent Dutch mathematician, physicist and astronomer, was already working in Paris as the Director of the recently established *Académie Royale des Science.*

The New Science Institutes: The Academies

Cassini, as previously mentioned, was also a member of the *Accademia del Cimento* in Florence, and the Grand Duke of Tuscany wished to have him permanently at his service, but the French Academy possessed a very different nature.

Historically, the 17th century witnessed the birth of several scientific academies, in the sense we currently define them as an official organization where science is conducted with the essential contribution of experimental verification. Small or large, in Italy or abroad, they included famous examples like the Royal Society in London, established in 1660, which is the most ancient of these societies that is still active. Founded in Colbert's house between June and December 1666, the *Académie Royale des Science*, or Royal Science Academy, whose history will be described in the next chapter, was however different from the others. It was not only granted a royal charter, like many other European academies of the time, but it was expressly founded as an organ of government under the authority of the Prime Minister, confirming the status of the monarchy.

The nominations of the academics and the orientation of research were strongly influenced by the decisions of Colbert and his successors, who commissioned the building of an astronomical observatory and funded several scientific expeditions around the world.

Since December 22, 1666, a small group of astronomers, mathematicians, physics, chemists, botanists, and zoologists have convened officially in the *Bibliothèque du Roi*, the King's Library in Vivienne Street, for twice-weekly working meetings (Fig. 12.3). For 30 years, such meetings took place on Wednesday for the "mathematicians", namely scientists active in the fields of physics and astronomy, and on Saturday for the "physicist", a definition that, at the time, included chemists, zoologists, and botanists.

Cassini agreed to join the French Academy, and he sent as his first scientific contribution a work about the observation of the lunar eclipse on May 26, 1668. He

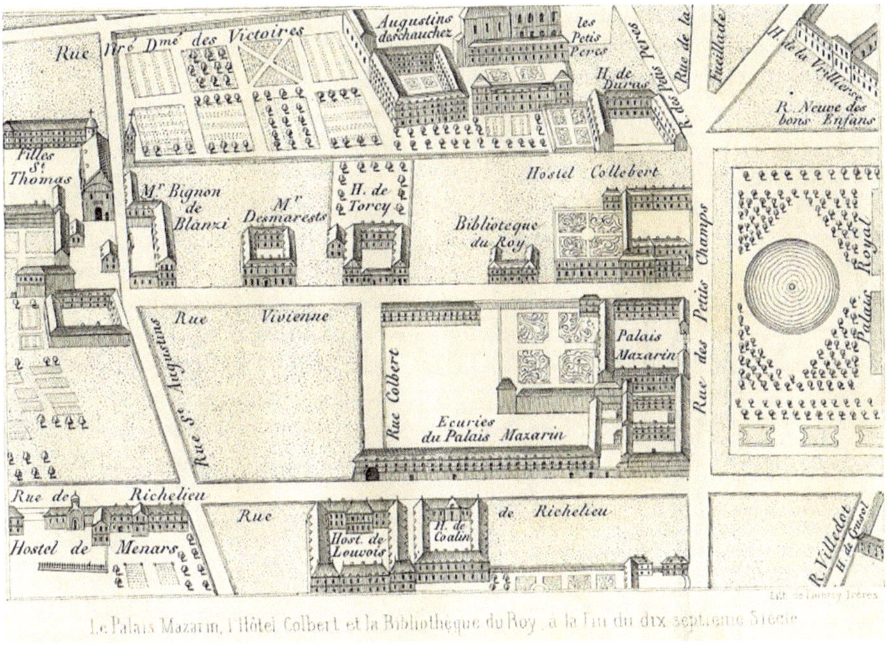

Fig. 12.3 Vivienne Street and the King's Library, 1845 (lithograph)

was in Rome in the Palace of the Cardinal of Este in the presence of scholars and nobility, and on such an occasion he also observed Mars, Saturn, and its rings.

For this observation of the lunar eclipse on May 26, 1668, Cassini used a clock and a set of various telescopes. In particular, he observed several phases and used the motion of several spots he had detected on the surface of the Moon to estimate the difference in longitude between Rome and Paris. Indeed, the instant in which such spots entered into the shadow of the eclipse were measured and compared with those observed in Montmartre by the members of the French Academy, similarly to the method of the satellites of Jupiter. The longitudinal difference was about 41 min of time, and these results were published in the *Journal des Savants* on July 30, 1668.

In the "Journal" publication, Cassini also included reports about the variation of the apparent diameter of the Moon, which he used to sketch a theory about the Moon's libration a few years later. Moreover, in addition to the report of the lunar eclipse, Cassini sent to the Academy also his latest and recently published work, the "*Ephemerides*", which was particularly appreciated by Abbot Jean Picard, who had been engaged for a long time in geodetic measurements.

In the same year, during travel in Tuscany, Cassini received another invitation from Louis XIV, something that changed his life forever. The king was inviting the Italian scientist to Paris, initially, for a short period, to plan the construction of the new astronomical observatory of the *Académie Royale des Science*, and the new Pope granted his consent for this trip.

Indeed, the Pope maintained a good relationship with France, and it was almost impossible to deny a favor to the powerful King. Thus, the two parties initially agreed on a stay of just eight or nine months, so that Cassini could return in time for the beginning of the new academic year.

It was clear, however, that this would not be just a short-term transfer. Colbert wanted Cassini in Paris, and he immediately began to inquire about his salary. Soon after, various diplomatic negotiations were set into motion. The result was a settlement in which the scientist could keep his chair at the University of Bologna indefinitely, with the guarantee to have it back upon his return.

The travel started on February 25, 1669, from Rome, and Cassini stopped in Bologna, Modena, and Genoa to pay visits to various friends. Then he made a stopover also at Perinaldo, to meet his parents, and, after a few days, he set sail for Paris. Even if he did not know it yet, he was saying a last goodbye to Italy.

Curiosity

Cassini's salary at the University of Bologna was 3800 "lire bolognesi", the highest among his colleagues. In addition, he earned 2400 more as an engineer at Urbano Fort, and a further 1800 for the "Water of the Chiane" and for the "Water of Bologna and Ferrara", a total of 8000 lire. The Prime Minister Colbert sent Cassini a sum of 1000 "*écus*" for travel expenses and an annual pension of 9000 "*livres*" while staying in France.

The project of the new Observatory of Paris was shown for the first time to Cassini by a member of the Academy he had met in Florence. The astronomer, for several reasons, had negative feelings about the building as it appeared in the drawings. The main one was, according to Cassini, that the project did not take into account the technical requirements of astronomy. This impression was then confirmed when he arrived in Paris and he could see better sketches. His opinions would cause him many conflicts.

Part III
Paris

Cassini left Italy for Paris on February 25, 1669, with the important assignment of building the new Astronomical Observatory. Here he would find King Louis XIV, the Prime Minister Colbert, and the Royal Academy of Science. His initial consultation for the new building, however, would turn into a permanent stay (Fig. III.1).

Fig. III.1 Ancient map of Paris, 1615

Chapter 13
Louis XIV

Jean-Dominique Cassini, mathématicien
From the list of workers remaining in the Louvre Palace

With a sum of 1000 écus, or crowns, for travel expenses, Cassini arrived in Paris, on April 4, 1669, after a forty-day journey, with stopovers in many towns for seeing relatives and colleagues. He was accompanied by Count Ercole Zani, one of his disciples from Bologna, who later continued his studies in England.

When Cassini arrived in Paris the construction of the Observatory had already reached the first floor, but it was not habitable yet, so he went immediately to the Royal Library, the headquarters of the Royal Academy, for his accommodation, but its apartments were already taken. Part of them were occupied by the librarian and mathematician Pierre de Carcavy and his family, while the remaining one had been occupied for three years by Christiaan Huyghens (1629–1695), distinguished scientist and president of the Academy (Fig. 13.2).

Instead, Cassini was lodged in the *Galeries du Louvre*, the Royal Palace, whose façade was designed by the same architect as of the Observatory, Claude Perrault (1613–1688). There he was in good company: several scientists and artists were staying in that place and were exempted from taxes, as a sign of good will to those who had distinguished themselves professionally.

G. Bernardi, *Giovanni Domenico Cassini*, Springer Biographies,
DOI 10.1007/978-3-319-63468-5_13

Fig. 13.1 Louis XIV, King of France, portrait by Hyacinthe Rigaud that was used as a model for the famous painting in the Louvre (1701, musée Condé, CC0 1.0 Universal Public Domain Dedication)

Fig. 13.2 Louis XIV visiting the Royal Academy of Sciences. Etching by Sébastien Leclerc I (1671). In the background is the new astronomical observatory [The Metropolitan Museum of Art (CC0)]

On April 6, Cassini was presented to King Louis XIV (Fig. 13.1) by the Prime Minister Colbert. Like the Observatory, Versailles Palace was still under construction, so the meeting took place at the Tuileries Palace. Cassini remembered very well this first meeting and the positive feelings he had of the King. Apparently, the Sun King was convinced that the advancement of science, and not only humanistic studies or the force of arms, was essential to establish a flourishing and renowned France. Cassini also remembered very well the courtesy of the sovereign. As he writes in his memoirs: "*Je me trouvé si flatté des bontés de S.M. et de la manière dont elle me traita, que je songeai plus dès-lors à mon retour en Italie…*" [I was so flattered by the courtesies of his Majesty and by the way he treated me that from that very moment I did not think to return to Italy anymore].

The French "Académie des Sciences"

At this point, it is necessary to give a more detailed description of the "Académie des sciences" or the French Academy of Science. This is of the oldest among the European academies, established just a few years after the Royal Society of London by King Louis XIV, at the suggestion of the Prime Minister Jean-Baptiste Colbert to create a general academy (Fig 13.2).

As reported very well on the web site of the Academy (http://www.academie-sciences.fr/en/Histoire-de-l-Academie-des-sciences/history-of-the-french-academie-des-sciences.html, site consulted on April 18, 2017):

> This was in keeping with the heritage of various groups of scholars who, during the 17th Century, met around a patron or a learned personality. Colbert chose a small group of scholars who met on December 22, 1666 in the King Louis XIV's library, recently installed in rue Vivienne, Paris and where they then held twice-weekly meetings.
>
> The first 30 years of the Academy's existence were relatively informal, since no statutes had been established for the institution. It was expected to remain apolitical, and to avoid discussion of religious and social issues, but in contrast to its British counterpart, the Academy was founded as an organ of government. On 20 January 1699, Louis XIV gave the Company its first rules. The Academy received the title of Royal Academy of Sciences and was installed in the Louvre in Paris. Comprising 70 members, in the 18th Century it contributed to the scientific movement of its time through its publications and played a role as counsellor to those in power. On 8 August 1793, the Convention abolished all the academies.
>
> Two years later, on August 22, 1795, the National Institute of Sciences and Arts was created, bringing together the former academies of the sciences, literature and arts. The Institute's first so-called 'class' or division (physical sciences and mathematics) was the largest (66 out of 144 Members). In 1805, the National Institute of Sciences and Arts was transferred to the premises of the College of the Four Nations. In 1816, the Academie des Sciences became autonomous, while forming part of the Institute of France; the head of State remained [and still is] its patron.

In 1835, under the influence of François Arago, the Comptes Rendus [Proceedings] of the Academy of Sciences were inaugurated and became an instrument of prime importance for the dissemination of French and foreign scientific studies.

At the beginning of the 20th Century, the Academy observed a decline in its activities and influence already been triggered by demographic reasons. In the face of accelerated development of scientific research in France and in order to remain faithful to its calling, the Academy adapted its structures and its missions. Indeed, it recently initiated a deep-reaching reform of its statutes, concerning Membership and Missions.

French Language

Despite the above-mentioned Cassini's intentions to settle permanently in Paris, it has to be remembered that the agreement was for a consulting assignment lasting a maximum of eight or nine months. Indeed, he had left his house, his astronomical documents, and his servants under the leadership of his Bolognese friend Monti. Moreover, there immediately arose a problem with the language, maybe the only one that could have convinced the Italian astronomer to step back from his initial resolution.

When he was presented to the King, he spoke in Italian because his French was poor, and fortunately the King and the Prime Minister were talking in Italian which, at that time, was still a widely-used language in the diplomatic environment. On this point, Count Gratiani, envoy of the Duke of Modena, advised him not to speak and write in French, ever, but this contrasted with the practice of the Academy. During the meetings, in fact, the use of French was strongly encouraged because it was the official language of the society.

He wrote in Latin his first report in Paris, which was about observations of sunspots, but the publications of the Academy had to be in French, so it was then translated by the librarian Carcavy. Unfortunately, Cassini found this translation unsatisfactory, and so he decided to write the next in French, requesting the needed revision and corrections by the abbot Gallois, Secretary of the Academy.

The reason of these difficulties is very simple: Cassini used Latin at the University of Bologna and in any scientific environment, as well as in his mail exchanges with non-Italian-speaking colleagues. This was because, similar to the English these days, Latin was the official language of the learned. Local languages were used only rarely by scientists, in Italy the most famous example being probably that of Galileo Galilei. In Paris, the Academy would have never allowed the use of another language, other than French, so Cassini was forced to learn French well, a task that was not simple at all, according to his words: "*J'avoue que cela me coûta beaucoup dans le commencement*" [I confess that it costed me a lot in the beginning].

Nevertheless, yet again in his notes, Cassini remembers that he did his best to cope with the language barrier and to reach a satisfactory knowledge of French. Apparently, he was successful enough in his attempts that, after just a few months, the King congratulated him for his progress during a meeting at the Observatory. Whether these compliments were just formal or true, we will never know, but Cassini did definitely do his best, and it is interesting to read an account of a curious incident on this regard that happened when he had just arrived in Paris.

Curiosity

The language problem was apparent from the very beginning of his permanence in France, on the occasion of the evaluation of the project of the new Observatory. Cassini was puzzled and expressed his criticism in front of the King, the Prime Minister, and the architect Perrault, who obviously did not like the comments and responds: "*Sire, ce baragouiner là ne sçait ce qu'il dit*" [Sire, this stammering guy here doesn't know what he says].

These words and the whole event are reported in detail in a publication by Giorgio Abetti ("*Gian Domenico Cassini e i Cassini*", Celebrazioni Liguri, vol. I, Urbino 1939) with the words of the great-grandson of the astronomer, Jean-Dominique Cassini (or Cassini IV). According to his account, Cassini had just arrived in Paris and was in the presence of the sovereign Louis XIV, who had ordered him to bring the plans of the Observatory and to hear the opinion of the scientist. Cassini, in bad French, began discussing with the architect, who defended his work with elaborate sentences. Cassini, instead, probably frustrated by the difficulties with the language in his attempts to explain his astronomical reasons, just overheated, with even worse results. It is quite possible that, because of his poor French and his excitement, the speaking of the astronomer became so uncertain as to sound irritating to the ears of the king and the prime minister.

It is at this point that the architect, definitely irritated, answered with the cited phrase: "*Sire, this stammering guy here doesn't know what he says*" to which Cassini "*wisely remained silent, but the King accepted Perrault's argument, and so was wrong*". The reasons for the concerns about the project of the Observatory will be analyzed soon in more detail, but they can be already understood from the final words of Cassini IV: "*Le plus habile architecte, s'il n'a point pratiqué l'astronomie, ne saura jamais construire un bon observatoire*" [The most expert architect, if he does not know astronomy, will never be able to build a good observatory]. Given these comments, it has to be stressed that, probably, even if Cassini could speak better French, the result would have been the same.

The dispute between the architect Perrault and Cassini was a tidbit, which quickly spread beyond the Court and among the colleagues, even those not belonging to the Academy. This is evidenced by the exchange of letters among scientists, like for example those between the mathematician Adrien Auzout (we will meet again soon) and the famous physician Marcello Malpighi, "father of microscopical anatomy, histology, physiology, and embryology".

Despite this unfortunate beginning, the relationship with the King was always of reciprocal esteem and appreciation. In the memoirs of Cassini, it appears that Louis XIV loved hearing about astronomical observations, and he regularly gave him appointments in his office. Here he enjoyed Cassini's explanations for quite a long time, sometimes together with the Queen, but also with Princes and Dukes who also acclaimed these astronomical speeches.

Claude Perrault, French architect and one of the first members of the French Academy of Science, was also a physician and anatomist. He wrote treatises on physics and natural history, but he died in 1688 because of an infection caused by a cut occurring during the dissection of a camel.

For its first three centuries, women were not allowed to be members of the Academy. This caused the exclusion of many women scientists, including the two-time Nobel Prize winner Marie Skłodowska Curie (1867–1934), the Nobel winner Irène Joliot-Curie (1897–1956), the mathematician Sophie Germain (1776–1831), and many other deserving women scientists. The Academy refused in particular Gabrielle Émilie du Châtelet (1706–1749), natural philosopher, whose most recognized achievement is her translation and commentary on Isaac Newton's book *Philosophiae Naturalis Principia Mathematica* [Mathematical Principles of Natural Philosophy]. The first woman admitted as a correspondent member was a student of the Curies, Marguerite Perey (1909–1975), in 1962. In 2016, the Academy welcomed only 28 women out of 263 members.

Chapter 14
The New Observatory

> *...car j'aurais voulu que le batiment même de l'Observatoire eut été un grand instrument...*
> [...because I wanted that the same Observatory building was a great instrument...]
> Giovanni Domenico Cassini's letter to Viviani, 6 July 1666

The conversation with the sovereign and the architect of the Observatory happened when Cassini had just arrived in Paris and could not communicate properly in French. His little success in expressing his doubts, therefore, should not be very surprising. His point, however, was quite simple and clear: the Observatory of the Science Academy was a majestic building, more apt to convey the grandeur of the French nation, than to achieve its nominal purpose, since its architecture was rather an obstacle to efficient astronomical work (Fig. 14.1).

The story of this building began two years before the arrival of Cassini, when the name of the two hectares close to Saint-Jacques was still "Le Gran Regard" or *the big look*. In the maps of Paris of the time, this place stood on the outskirts of the town and was bordered on three sides by the gardens of three convents. To the north, the Abbey of *Port-Royal de Paris*, to the east the Capuchin novitiate, and to the west the novitiate of the Fathers of the Oratory, while the south side was just countryside with windmills.

On March 7, 1667, this land was bought by Minister Colbert on behalf of the King to build the headquarters of the recently established Royal Academy. The place would be used for the meetings of the society, for physics experiments, but especially for astronomical observations.

The Astronomical Observatory was officially inaugurated on Tuesday, June 21, 1667, the chosen day was, intentionally, the summer solstice, and in the minutes of the Academy it is reported that several academics, astronomers and mathematicians attended the event. "... *Auzout, Frénicle, Picard, Buot et Richer...*" gathered on the place where arose the Observatory "... *pour tracer une ligne méridienne sur une pierre que M. Couplet avait fait poser pour prendre la hauteur méridienne*", that is,

G. Bernardi, *Giovanni Domenico Cassini*, Springer Biographies,
DOI 10.1007/978-3-319-63468-5_14

Fig. 14.1 The Paris Observatory in the beginning of the eighteenth century. The tower on the right was the wooden Marly Tower, originally built to lift water for the Versailles reservoirs and fountains, that astronomer Giovanni Domenico Cassini moved to the grounds at great expense for the mounting of long tubed telescopes and even longer tubeless aerial telescopes

to draw a meridian line on a stone which Mr. Couplet had ordered to put in place for the measure of the meridian height.

Actually, using two sextants they traced what will be known later as the Paris Meridian, namely the meridian line crossing the Observatory building and whose path is still marked today on its premises. After finishing this symbolic task, work immediately began by digging about 30 m deep into the ground. This work, eventually, ended in a grandiose building designed by the previously mentioned architect Perrault, also a member of the Academy. Unfortunately, the massive building was not optimized for astronomical observations. In particular, the visibility of the sky was often obstructed from many places, as Cassini attempted to explain in the presence of the King. His complaints however, as we have already revealed above, did not get a favorable reception because the poor French language skills of the Italian astronomer. With the construction already having arrived at the first floor, Cassini immediately suggested several changes, *...car j'aurais voulu que le batiment même de l'Observatoire eut été un grand instrument...* because he wanted that the same Observatory building to be a great instrument, as he wrote in one of his letters to the Italian colleague Viviani, but only a few changes were granted.

This recalls the previous project in Bologna of the meridian line in San Petronio where, as the intention of Cassini, the whole building was supposed to serve as an astronomical instrument. In this case, however, the scientist surely perceived the historic change in astronomy caused by the advent of the new lens-based instruments. The scientist had already witnessed such a change in Italy, and, indeed, he had been an active participant with the new lens manufacturers who would greatly increase the precision of the astronomical measurements. This change was gathering momentum in those years, but it was neglected by Perrault and also by the other members of the Academy. For example, for the determinations of the Observatory's coordinates, they used sextants without spyglasses, and only in the following year did they begin to mount them.

Cassini wrote these words about the modifications he had proposed: "*…je proposai d'abord qu'on n'élevât ces tours que jusqu'au second étage, et qu'au-dessus on bâtit une grande salle carrée, avec un corridor découvert tout à l'entour, pour l'usage dont je viens de parler. Je trouvais aussi que c'était une grande incommodité de n'avoir pas dans l'Observatoire une seule grande salle d'où l'on pût voir le ciel de tous cotés, de sorte qu'on n'y pouvait pas suivre d'un même lieu le cours entierdu soleil et des autres astres, d'orient en occident, ni les observer avec le même instrument sans le transporter d'une tour à l'autre.*" In other words, he proposed to cancel the making of the towers over the second floor, but rather to build above them a large square room with an uncovered hallway running all around. He also noted that it is very convenient to have a single large room to observe the sky in every direction so as to follow the Sun and the stars from east to west without having to move the instrument from one tower to the other (Fig. 14.2).

The little changes conceded, instead, were as follows: the octagonal north tower became a square tower with an opening at the top used for observations at zenith, sheltered from the wind, and in the middle of the building was authorized the construction of a 55-meter-deep well, reached by a spiral staircase down into the cellars. The corresponding vaults had holes for zenith-stars observation and to reduce the effect of atmospheric refraction. Moreover, the well was used for physics experiments, for example, on falling bodies or on the change of temperature at different heights.

When the Observatory building was finished, Cassini moved into his accommodation inside with Couplet, his young assistant. He had previously stayed in the "Galeries du Louvre" (today one of the most famous museums in the world), and his astronomical observations were made in the garden of the Louvre or the "Bibliothèque du Roy" (King's Library) in rue Vivienne where there was also the largest and most accurate solar quadrant in Paris. Cassini decided to rent a house with a garden near the West Gate of the city to Ville-l'Evèque to set his astronomical instruments such as his precious Campani lens.

Fig. 14.2 Portrait of Giovanni Domenico Cassini with the new Astronomical Observatory in Paris

Curiosity

The last lines written by Cassini in his memoirs regarded the Astronomical Observatory's construction and his negative opinion about its architectural design. They are fully reported in Appendix B, in French, but we can summarize the discussion here. Cassini immediately noted that the building was compact, with four main walls oriented along the cardinal axes, but with three prominent towers jutting out of the main body of the edifice. Two were placed on the eastern and western corners of the southern wall, and the other was in the middle of the northern one. These large structures seemed to hinder the astronomical activity that is the main purpose of the building (Fig. 14.3).

Fig. 14.3 The Astronomical Observatory of Paris today (Author's photo)

The reason can be easily understood by recalling the kind of astronomical instruments used at the time. Indeed, Cassini reveals in his comments that he wished to add four large quadrants with the accuracy of seconds of arc. To understand the ambition of such an instrument, it is worth recalling that Tycho Brahe, about 70 years later, had reached an accuracy of slightly less than 1 arcmin, that is 60 arcsec, and that Flamsteed would get to about 20 arcsec approximately by the beginning of the 18th century. Such an accuracy could be reached only by large mural quadrants, for which Cassini basically needed to exploit the large dimensions of the building. Actually, this was his attempt to use the entire building as a great astronomical instrument, but this far-reaching goal was made impossible because of the towers. In addition to this, he notes that their octagonal shape is not appropriate for astronomical uses, because the short sides leave only limited space for doors and windows. Unfortunately, it is not possible to compare his comments with the original project of Perrault because the original designs of the Astronomical Observatory have been lost.

"*Sic itur ad astra*" [So we go to the stars] is the Latin motto written on one side of the medal minted on the occasion by the Foundation of the Astronomical Observatory, while on the other side appears the image of Louis XIV. Following the tradition, a gold and a silver example of this medal were placed in the foundations of the building.

The two sextants used to trace the meridian line of Paris were without a spyglass but with optical sights. One, in iron and copper with a radius of six feet, was the property of the Academy, while the other was owned by Picard, one of the members of the Academy active in the fields of astronomy and geodesy. It was not uncommon at the time for the scientists to buy their own scientific instruments for their studies.

Chapter 15
Sunspots

Anno 1671, die 14 septembris,
Deo auspice, in regium observatorium veni,
ubi mihi parabantur cubicula,
qui plena erant fabris portas fenestrasque munientibus
Giovanni Domenico Cassini

Leafing through the first volume of the "*Journal des observations faites à l'Observatoir royal de Paris*" [Journal of observations from the royal observatory of Paris] one can find, just at the beginning, the previously cited Latin note of Cassini. They describe the settlement of the scientist at the Royal Observatory, and the translation reads: "In the year 1671, 14 September, with the help of God I came in the Royal Observatory, where some rooms were being prepared for me, which I found full of workmen intent on mounting doors and windows."

So in 1671 Cassini finally left his rented house in Ville-l'Evèque and moved with his optical instruments and his young assistant Couplet into the new Observatory. Although not yet finished, the architecture was changed not only to address, at least partially, the observational problem suggested by Cassini, but also to create an apartment for the astronomer on the first floor, which included the much-needed chimneys for heating and which had not been considered in the original project.

Here, Cassini wrote a booklet with specific guidelines for the observational activity entitled "*Istructions à ceux qui travaillent à l'Observatoire*" [Instructions to those who work at the Observatory] in which he summarizes the observations to be done on a daily basis. They included the measurement of the times of the rise and setting of the Sun and of its apparent diameter. The same measurements had to be done for the Moon and the planets, to determine the refractions and parallaxes near the horizon. Special attention was given to the brilliant stars (first magnitude) crossing the meridian, useful for the corrections of the astronomical tables. Meteorology was also considered as temperature, barometric pressure, and wind direction had to be recorded.

G. Bernardi, *Giovanni Domenico Cassini*, Springer Biographies,
DOI 10.1007/978-3-319-63468-5_15

Today we just assume that any scientific program is based on such a kind of systematic and organized work, but taking for granted that this was common practice in the past would underestimate the revolution represented by this approach for the scientific research of that time. Indeed, such a method was quite new and very rare at the beginning of the scientific revolution. With just some notable exceptions like that of Tycho Brahe, most of the astronomers observed sporadically, following neither long-term goals nor systematic-measurement campaigns.

His observations on French territory, however, had begun already before entering the Observatory, actually just upon arrival in Paris of the scientist. Cassini needed to find an accommodation suitable for his systematic-observation program, and he found it at Ville-l'Evènque (a suburb of Paris) in the garden of his temporary residence. Here he observed for the first time the sunspots, an event he remembered with these words: *J'aperçus là, pour la première fois, des taches dans le soleil don't je fis la description qui fut envoyée au Roi à Fontainebleau.* [Here I saw for the first time the sunspots, whose description I sent to the King, at Fontainebleau]. Unfortunately, the report of his observations was written in Latin, and thus it was not published. Consequently, as already mentioned, Cassini sent the manuscript to the Academy librarian Carcavy for translation into French. Dissatisfied by the result, he decided to translate personally his works in the future (Fig. 15.1).

His investigation of the sunspots, however, had at least one practical consequence. Following his usual methodical approach, in fact, Cassini conducted regular observations for several days, as he reports in his writings, which allowed him to calculate the speed of the apparent motion of the spots and to make predictions on their return to the same locations on the solar disk. According to his calculations, this event had a periodicity of 27 days, as seen from the Earth. Actually, this coincides with the so-called *synodic period* of rotation of the Sun, which is about two days longer than the sidereal period. The two in fact represent the rotation period with respect to the Earth and the star, respectively, and their difference is due to the motion of our planet around the Sun. The predictions of this theory was eventually confirmed, and because of this success, Colbert also wanted to watch the Sunspots from the garden. The Prime Minister was so pleased to see these new objects and to use the excellent lens by Campani, that he personally encouraged the transfer of Cassini to the new Observatory.

Actually, as remembered by Cassini, his transfer to the new location was dated September 14, almost simultaneous to the appearance of the sunspots beyond the visible side of the Sun as foreseen by his theory. But, previously another very important observational evening had happened: on September 7, 1671, Colbert, Carcavy, and Huygens (Fig. 15.2) with his telescope convened in the Academy Library, together with Cassini. The Prime Minister stressed that good telescopes were necessary to make new astronomical discoveries, and Carcavy (who was also a mathematician and scientist) confirmed that the Academy had tried for a long time to build powerful lenses, unfortunately with poor results. So the Prime Minister

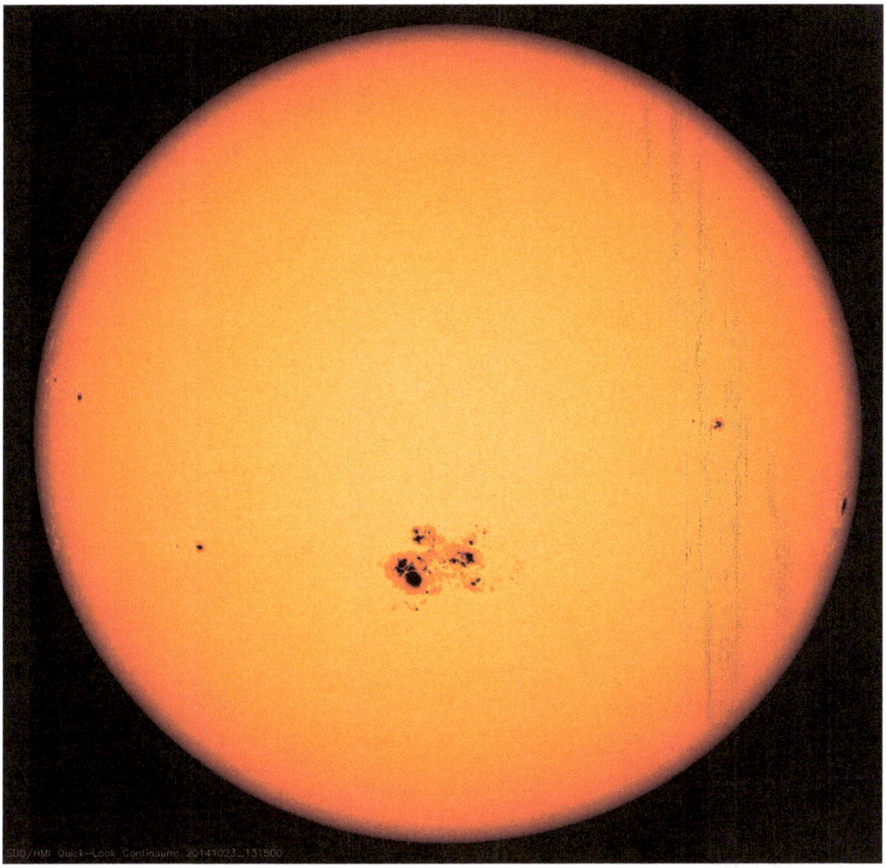

Fig. 15.1 Image of the sun's solar disk, showing sunspots on the surface in October 2014 (NASA/SDO, CC-BY-2.0)

inquired about Campani and his lenses, eventually resolving himself to order some of these high-quality optical tools funded directly by the King. Campani realized an excellent lens with a focal length of 34 ft (more than ten meters) about twice that of the lens used by Cassini for his observations of the sunspots.

Theory of Sunspots

Cassini was not only a skilled observer with the best equipment you could find on the market, but he had also great intuition. In the case of Sunspots, his predictions were not just the result of some calculations; rather, they were also based on a theory (see Appendix C for the original French exposition) that explains in detail

Fig. 15.2 Christiaan Huygens. Line engraving by F. Ottens (Wellcome Library, London (CC BY 4.0)

the motion of these new objects. First of all, Cassini, presumes that these spots are on the spherical surface of the Sun and he establishes that they describe parallel circles around the two poles, whose axis forms an angle of about seven and a half degrees with respect to the ecliptic. The sunspots make their revolutions in a time nearly equal to that of the periodical revolution of the Moon around the Earth, however Cassini complains that they are not always visible by telescopes on their return.

To explain this, he supposed that the Sun, like the Earth, is composed of two substances: a solid and a liquid one, similarly to the continents and the seas on our planet. The solid one would be opaque, and the liquid would be the light-emitting matter, which covers the greater part of the opaque matter, leaving only some isolated "rocky" places more opaque than the rest of the surface, thus appearing as spots. Then Cassini also supposed a kind of tidal movement, similar to that of our seas; the ebb and the flow periodically raises and lowers the luminous matter, thus varying the dimensions and the shapes of the emerging opaque parts, and therefore the appearance of the spots. Although nowadays this theory can seem quite strange, we have to remember that the real nature of the solar composition started to be known only in the 19th century, with the discovery of helium, and the explanation of the tides based on the gravity force will be first given by Newton in his *Principia*, published in 1687.

As reported in his memoirs: "*Those which we observed in the beginning formed at first in the shape of a scorpion with its feet and its tail. A little later, this part detached and formed smaller spots separated from each other. They were enveloped in a kind of nebulosity, which represented to our imagination the whirlwinds which form around the points of rocks by the tides. It could also be supposed that, as in the globe of the earth there are volcanoes, which at certain times throw flames and ashes round them, so there were in the sun.*"

The volcanoes hypothesis concluding his theory had been motivated on the basis of some apparently recurring spots. Actually, Cassini claimed to have observed that, in several cases, a spot appeared again and again in the same position on the surface of the Sun, determined with respect to its poles. Since the movement of a liquid cannot be so regular, he supposed instead that some sort of solar volcanoes emerged from the surface, and periodically signaled their presence with this activity, perceived by the observers on Earth as spots. To better explain this position, he compares this situation to how an observer far from the Earth would see Vesuvius. It would be visible only during its active periods, appearing always in the same position with respect to the poles of the Earth. Cassini assumes that his conjectures are as probable as those of the return of the same planets to the same place in the sky after a number of revolutions, and he recounts, for example, the case of Mercury or of Phosphorus and Hesperus, which formerly were supposed to be two different stars, but were finally recognized as the same planet Venus.

Curiosity

The large palace built for the Academy of Sciences and its members would host, at the behest of Minister Colbert, a Museum of Anatomy and a collection of curiosities of natural history, in addition to the labs for chemistry, physics, and mechanics. Unfortunately, the building was unsuitable for many uses, and the simultaneous presence of several scholars during the experiments was more distracting than useful. In addition, because the building was in the suburbs and far from the center, academics continued to meet in the King's Library, the original headquarters.

Therefore, its rooms became warehouses for the machines and models built by scientist, and it was eventually used only as an astronomical observatory, especially because of the influence of Cassini. Indeed, his presence in this place is attested between 1671 and his death, in 1712, by the observational logs called: "*Journal des observations faites à l'Observatoir royal de Paris*" [Journal of observations from the royal observatory of Paris]. Daily observations with notes were recorded by Cassini in Latin until 1683. Afterwards, he would always use French, his newly adopted language. Mysteriously, the logbooks dating between 1674 and 1680 are lost, and already Cassini IV, his descendant, in 1770, decried this gap.

Some observers considered the sunspots as planets and gave them the name of *Sydera Borbonia*.

The observational logbooks are mainly "in folio", namely folded only once, while those of the last years are "in quarto" (folded twice) and just one is in "octavo" (folded three times). There Cassini records not only the data from the observations, but often anecdotes, drawing, thoughts, and weather data such as temperature and atmospheric pressure, or he even inserts additional pages with more calculations. In this way, this documentation not only provides scientific data, but aspects of the personality of the scientist. It also reports important unconfirmed discoveries. For example, in October 1671, a few months after his entry in the Observatory, he drew a little star with a tiny Latin annotation: "*poterit esse Supremus Satelles*" [may be the above satellite], where Cassini had already realized that he was observing a newly discovered satellite of Saturn.

Chapter 16
Saturn

Je m'attachai ensuite à observer Saturne...
Giovanni Domenico Cassini

Even if the notes dictated by Cassini to his secretary were coming to the end, there are several remarkable astronomical works and discoveries that have still to be unveiled. As referred to in the previous chapter, the sunspots and, in particular, the observations made together with Jean-Baptiste Colbert became the key that would literally unlock the doors of the still unfinished observatory of Paris to the astronomer, who would finally occupy his apartment on September 14, 1671. This date is fixed in the memory of the Ligurian scientist because it was almost coincident with the appearance of a series of spots on the visible side of the solar disk, as foreseen by his theory.

Here, Cassini installed a program of regular astronomical observations that quickly made the Paris observatory the most important in the world. Cassini also began to observe Saturn, and, as mentioned in the previous chapter, during the night of October 25, 1671, he spotted for the first time another satellite of this planet. Currently known by the name of Iapetus, initially it was called "Supremus", possibly because of its great distance from the central body. Meanwhile, another foreign member of the Academy, actually its president, the mathematician Christiaan Huygens, was observing Saturn with good results.

In 1655, in fact, with a refracting telescope designed by himself which featured a magnifying power of 50, Huygens had discovered the first and largest of the moons of Saturn: Titan. Moreover, he had also observed the rings of Saturn, thin and flat, and made the first estimation of their inclination with respect to the orbit of Saturn. The value of 23.5° determined by the Dutch scientist was later emended by Cassini to 34°. This correction was not a cause for dispute between the two scholars, as noted by Cassini: "*Huyguens n'en fut pas fàché et vint plusieurs fois assister à mes observations.*" [Huygens was not at angry and came to my observations several times]. But how much these two, very important scientists of the French capital were different, or similar?

G. Bernardi, *Giovanni Domenico Cassini*, Springer Biographies,
DOI 10.1007/978-3-319-63468-5_16

Huygens and Cassini actually had very different personalities: Christiaan Huygens, in a nutshell, was a solitary genius, as reported by Albert van Helden: "*Intellectually he was a loner, a scientist who was happiest in his study with pen, paper, and a few instruments. There has, of course, always been a need for such scientists. To make a public career in science in the seventeenth century, however, one had to sustain a public posture, and I think that this was uncomfortable for Huygens all his life. The public life took away too much energy from his beloved private study. And to be a drudge at the eyepiece of a telescope every night for one's entire life was anathema. When he was on the trail of something important, he observed assiduously, but after he had satisfied himself he stopped and turned back to mathematical studies, always the center of his interest.*"

Another important difference was their families. Huygens came from a very wealthy and influential family, while Cassini, as we learned at the beginning of this book, was not poor, and even well-fixed, but certainly of a much lower level of wealth. Huygens refused a career as diplomatic advisor, like his father, because it bored him to tears; on the contrary Cassini acquired the manners of the courtier and was very successful in this role. The academic career pursued by the latter was considered prestigious and a natural way to increase his social status, given his origins, while, for a person of the social level of Huygens, this would have been considered diminishing, and only an appointment as the president of the Academy could be acceptable.

Huygens was among the finest scientific minds in Europe and also a significant inventor; he made a revolutionary new clock that improved the accuracy of timepieces by orders of magnitude. Cassini, on the other hand, was an expert observer and an unparalleled manager and organizer, who first established the procedures of modern observatories and whose view of scientific work was, in this sense, much more advanced than others among his contemporaries.

Was there competition between them? According to van Helden: "*A few years after Cassini's arrival in Paris, Huygens remarked to his brother Constantijn that Cassini was at the telescope every clear night and that he [Huygens] would never want to do that, being satisfied with his earlier astronomical discoveries which, at any rate, were much more important than those of Cassini's.*" Huygens thought that his mathematical constructions had to be perfect and elegant and so published usually in a slow manner, so that much of his scientific work and discoveries remained unpublished at the time of his death. In sharp contrast, Cassini always published his results quickly and left virtually no unpublished works at his death.

The Aerial Telescope

The number 34 returns in the history of Cassini's exploration of Saturn with a new and powerful telescope having a focal length of 34 ft, that is, more than 10 m. It had been commissioned by Colbert and King Louis XIV, and made, as usual, by Campani. This was the instrument that Cassini used to continue his observations of the planet and with which he discovered, on December 23, 1672, very close to Saturn, a third celestial body revolving around Saturn. Again, because of its position with respect to the planet, the astronomer called his second discovery "*Intimus*", nowadays known as Rhea. The orbit of these two new satellites were described by Cassini in a volume published in 1673 with a fawning dedication to Louis XIV. Many colleagues criticized the obsequious "*astronome mondaine*" [socialite astronomer] Cassini, but it is also true that others acknowledged his availability and generosity.

This new generation telescope was so long that it could not be used from inside the observatory building, thus external structures like the wood tower shown on Fig. 14.1 were set up to enable its use from the outside. This tower was called "the Marly", because it had previously been used in Versailles as part of the so-called *Machine de Marly* to lift water for the reservoirs and fountains at the gardens, and now it was moved from there to the Observatory grounds to serve as a place for attaching the lenses of aerial telescopes. But what is an aerial telescope? It is a type of refracting telescope with a very long focal length that did not use a tube, so that the light thus travels from one lens to the other through the open air (Fig. 16.1).

The idea may have emanated among the astronomers from Christiaan Huygens and his elder brother Constantijn, but it is not clear if they actually invented it. What is actually known is that this kind of instrument was built in the second half of the 17th century as a first way to meet the need for more powerful telescopes. One of the things that people tried to achieve was greater magnification which, with a given ocular lens, could be obtained by increasing the focal length of the lens at the other side of the optical system. This obviously called for longer and longer telescopes, which were difficult to handle also because of the weight of their structures, and the absence of a tube was certainly an advantage in this sense. However, handling an aerial telescope was still not an easy task at all, and it seems that only Huygens and Cassini were able to get good results from them. The objective lens was mounted on a structure, as simple as a pole or a tree, or in a more complex manner such as the tower as in the case of the observatory of Paris, on a swivel ball-joint, while the observer stood on the ground and held the eyepiece which was connected to the objective by a string or a rigid rod; by holding the string tight and maneuvering the eyepiece, the observer could aim the telescope at objects in the sky. Actually, it was also because of the extreme difficulty of using these very long focal-length telescopes (Fig. 16.2) that led astronomers to develop alternative designs like the reflecting telescope.

Anyway, Cassini made some personal modifications to improve the usability of this instrument. For example, he devised what he called a "*tube à trois faces d'échelles*" which is described in the book: "*L'opera del genio italiano all'estero*" by Savorgnan di Brazzà. Literally, the French expression means "three-sided ladder tube" and it was a sort of 20-m "tube" shaped as a triangular prism. Its wooden faces were not solid, but the rigidity of the structure was improved with respect to the standard rod thanks to a sort of "ladder-like" framework. The same arrangement included a mechanism that allowed the eyepiece to be moved back and forth along

Fig. 16.1 A large refracting telescope, in use outdoors [Wellcome Library, London (CC BY 4.0)]

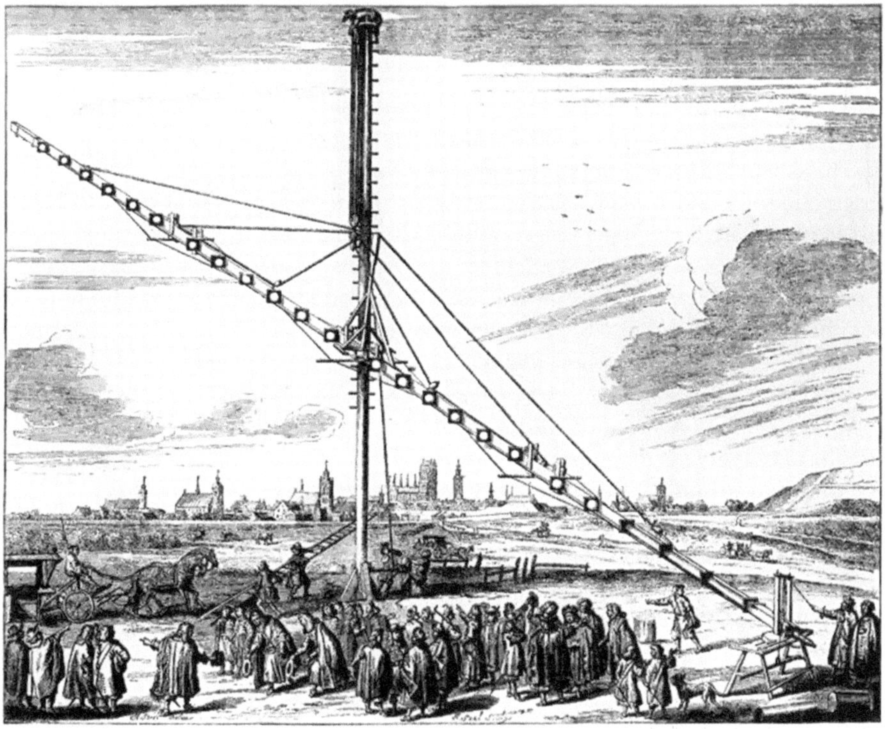

Fig. 16.2 Woodcut illustration of a 45-m-long astronomical refracting telescope built by Johannes Hevelius

some grooves, and thus to set the focus with respect to the lens on the opposite side, which instead was fixed.

Among other things, Giovanni Cassini elaborated a mount for spyglasses or telescopes with a clockwork system called "*machine parallatique*" or parallax machine that, similar to modern instrumentation, could compensate for the Earth's rotation, so that one could fix a star in the sky and follow it for the whole night.

Saturn, Satellites and Rings

At the end of his memoirs, Cassini recalled the nights of the discovery of his first satellite, when he realized that it was a satellite rather than a star, and it is interesting to read his writing and notice his approach: "*Saturn était alors avec le petites étoiles de l'eau d'Aquarius, avec lesquelles je le comparais autant de fois que la sérénité du ciel me le permettait, en faisant souvent la descriptions de la configuration de ces étoiles avec Saturne et avec l'ancien satellite. Je la trouvai un peu variable d'un temps à l'autre, et je vis dans la suite que cette variation pouvait s'attribuer à l'étoile que j'appelai alor second satellite, parce que ce fut le second de ceux qui furent d'abord découverts*".

To sum up, when Cassini observed Saturn, the planet had the stars of the Aquarius constellation in the background. He recorded the configuration of this system as many times as possible, whenever the sky was clear enough to allow its observation, monitoring the relative positions of these stars with respect to Saturn and the first satellite, Titan, discovered by Huygens. Cassini observed that such a configuration varied slightly over time, and he later realized that these variations could be attributed to an object that he had identified as a star, understanding that it was the second satellite of Saturn (Rhea) instead.

Après que j'eus observe en quelque manière la vitesse apparente de son mouvement particulier, j'invitai l'Académie à venir l'observer. On trouva une petite étoile à l'occident de Saturne, où j'avais dit que ce satellite devait paraître, et pour lors la compagnie fut satisfaite et commença à calculer la période de sa révolution, par la comparaison de mes premières observations avec cette dernière.

Cassini observed the apparent speed and the details of the motion of this new astronomical object and afterwards he invited the members of the Academy to come and observe it as well. They found it to the west of Saturn, where the discoverer had told them to look at, and then they began to calculate the period of its revolution by comparing their observations with the first ones done by Cassini.

Mais je n'en fus pas satisfait, parce que je ne trouvai pas cette étoile assez précisément à la distance de Saturne qu'elle devait avoir par mes premières observations; et quand M. Huguens se hasarda de determiner sa revolution qu'il donna dans un écrit cachet au secrétaire, je lui dis que nous n'avions pas ancore toutes les observations nécessaires pour cette determination.

But Cassini was not satisfied because he did not find the satellite precisely where he would have expected from his first observations, and he stopped Huygens, who had tried first to determine the orbit, understanding that they needed more observations (Fig. 16.3).

Il était arrive à cette planète un accident qu'on n'avait pas ancore observé dans les autres: c'est qu'elle n'est pas visible dans toutes les parties de sa revolution qui

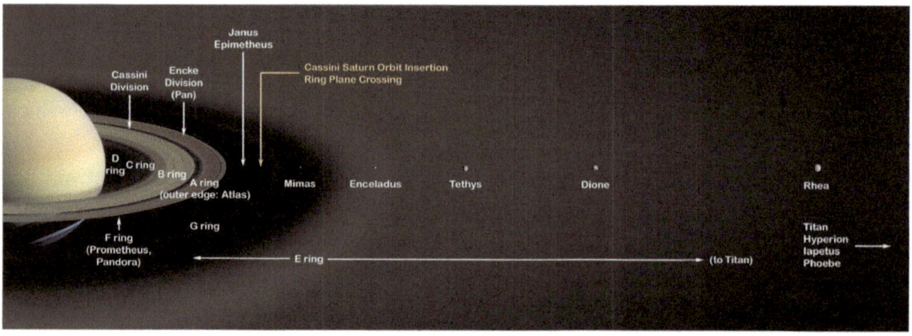

Fig. 16.3 Artist's concept of Saturn's rings and its major icy moons (NASA/JPL)

sont à pareille distance de Saturne, ni dans toutes ses revolutions; ce qui peut s'expliquer par des taches places à sa surface, et par la revolution autour de son axe, par laquelle elle tourne à la terre tantôt la face tachée, tantôt celle qui ne l'est pas, ou à quelqu'autre cause physique qui rend ces entroits du satellites tantôt éclatans comme il arrive aux volcans de la terre, et tantôt sans cet éclat. En effet, ce satellite fit plusieurs revolutions sans paraître, et parut ensuite long-tems. Cette alternative a lieu encore aujourd'hui; ce qui n'a pas empêché que par les observations qu'on en a pu faire, on ait appris la période de sa revolution, qui est de 79 jours 22 heures.

Cassini reports that it is quite difficult to observe this satellite. The main problem is that it is not always visible during its orbits. Its appearances are quite irregular, and it can even remain invisible for many revolutions and then stay visible for a long time. In the end he was able to estimate a period or revolution of 79 days and 22 h.

Je découvris ensuite en divers autres tems trois autres satellites de Saturne, don't j'ai parlé dans divers Mémoires que j'ai publiés, et dans les registres de l'Académie des Science. M. Duhamel, qui était alors secrétaire, en a aussi parlé dans l'histoire latine de la même Académie.

With these last lines to his secretary mentioning the discovery the three more satellites of Saturn, Cassini, by this time blind, concluded his astronomical memoirs. He stresses the fact that this achievement was published not only in the Science Academy journal, but also in various essays like, for example, in the work about the history of the Academy written by the secretary Duhamel.

Cassini, unfortunately, did not have the time to tell about, in his memoirs, his observations of the rings of Saturn and all the work that he did later. As previously recalled, the discovery of the rings of Saturn has to be ascribed to Huygens, who recognized their true shape in 1655, but in the "*Journal des Sçavans*" of March 1, 1677, Cassini reported a particular detail about his observation of the rings: "*...la largeur de l'Anneau étoit divisée par une ligne obscure en deux parties égales, dont l'intérieure et plus proche du globe étoit fort claire, et l'extérieure un peu obscure. Il y avoit entre les couleurs des ces deux parties à peu près la même difference qui est entre l'argent Mat, et l'argent bruni...*".

For the first time the ring is recognized as a composite structure, divided in two equal parts by a dark line that separates an internal ring, closer to the planet and brighter, and an external one, which was darker. It is because of this description that today we know this gap as the *Cassini Division*. Later, the difference between the colors of these two rings is described as being almost the same as that between natural and burnished silver.

Actually, the hypothesis that the ring surrounding Saturn was not homogenous had already been made by Campani. In a letter sent to Cassini in 1664 and kept in the Vatican Library (Chigi III 462) where he writes: "*...il Cerchio per la parte di fuori, cioè verso la circonferenza esteriore essere men chiaro e lucido per fino alla metà del suo piano; e dalla metà in là verso il globo di Saturno esser più chiaro e luminoso del medesimo globo...*" [...the external part of the Circle (the ring of Saturn) that is towards the outer circumference to be less bright and shiny up to half of its surface; and from the middle towards the sphere of Saturn to be clearer and brighter than the planet itself…]

The astronomer, thanks to the instrumentation of Campani and to his intuition, makes a step further in understanding the nature of the rings with these important words: "*L'apparence de l'anneau est cause par un amas de très petits satellites de différents mouvements qu'on ne voit point séparément...*" in other words, he comes to the modern view of the rings by stating that their appearance is the result of many small satellites moving independently of each other and cannot be seen separately. In the same journal, Cassini accompanied his statements with a famous engraving obtained from a drawing made in 1676 in which you can see Saturn and its ring depicted in two different colors. Surely this picture cannot compete with the details in the spectacular images of the rings made available by the space mission NASA-ESA-ASI Cassini-Huygens mission, but it is remarkable how, almost 350 years ago and with infinitely less sophisticated instrumentation, this great scientist could understand the actual nature of this majestic cosmic structure.

Even so, the adventure with Saturn does not end with the rings. In fact, by using Campani's lenses with a very long focal length set up on the grounds of the Observatory, in 1684, Cassini announces to the Academy the discovery, on the night of March 21, of two more satellites: Tethys and Dione. Galileo Galilei, calling them "the Medician stars", had dedicated the first four moons of Jupiter to his soon-to-be patron Cosimo de' Medici, the Grand Duke of Tuscany, and his family. Similarly, Cassini dedicated his first four satellites (all those known at those times except for Titan) to the French king, calling them *Sidera Lodovicea*, or the stars of Louis. For this discovery, he was awarded a medal by the King. On one side of the medal is written: "*Saturni satellites primum cogniti*", the Saturn satellites known for the first time.

Curiosity

Louis XIV paid to Campani the considerable sum of 1000 "scudi" or ecus, an old coin of the time, for the making of the powerful 34-ft telescope.

Titan, Iapetus, Rhea, Tethys, and Dione are the names of the first discovered satellites of Saturn; Titan, the first, was discovered in 1655 by Huygens, while the following four were discovered by Cassini in 1671, 1672, and 1684, respectively. The names, taken from mythological characters, were assigned by the astronomer John Herschel, son of William, in the 19th century. Cassini first used Latin names based on their positions: Supremus [above], Intimus [inner], Medius [middle], Proximus [next], and afterwards he dedicated all of them to the King of France using the name "Ludovici", from Ludovicus, the Latin translation of the King's name Louis.

The "*Cassini Division*", a 4800 km (3000 mi) wide gap between the "A Ring" and the "B Ring" was discovered by Cassini and is named after him.

Chapter 17
Jean Dominique Cassini

Moglie secondo il mio cuore

Jean Dominique Cassini

We have now to leave Saturn, its rings and its satellites and jump back in time to better understand the evolution of Cassini's settlement in France and how Giovanni Domenico Cassini became Jean-Dominique Cassini. In other words, how the Italian astronomer became a French astronomer.

Initially, he arrived in Paris in 1669 with the limited and very specific assignment of setting up the new Observatory. This was agreed to by the Pope, after a long and difficult diplomatic negotiation among the Vatican, the French Minister Colbert, and the University of Bologna. The latter, in particular, did not want to lose such a famous astronomer and its highest-paid university professor, so when Cassini left for Paris on February 25, 1669, he still maintained his chair at the University of Bologna. Indeed, the agreement was that, on his return, he would get back his place and assignment.

A few months after his arrival in Paris, Cassini wrote a letter to the University of Bologna to inform them about the tasks assigned to him by the King. Prefiguring negative reactions to his report, he pre-emptively reviewed the actual situation with the Apostolic Nuncio (the Pope's ambassador) in Paris, Monsignor Bargellini, and then with Camillo Paleotti, the Ambassador of the Senate of Bologna in Rome. While Cassini spent time in Paris, in Bologna discontent was spreading among his colleagues, and he also risked excommunication by the Pope. However, as recounted in previous chapters, he was enrolled in the "Rotuli" of the university, and this protected him from this procedure.

G. Bernardi, *Giovanni Domenico Cassini*, Springer Biographies,
DOI 10.1007/978-3-319-63468-5_17

Unfortunately for the University of Bologna, he never came back to his chair. His temporary stay in Paris would become permanent despite several reminders sent to him from Italy, as Anna Cassini reports in her book "*Gio: Domenico Cassini*". A thick mail exchange occurred between Bologna, Paris, and Rome, but the ties with the French capital strengthened further in November 1673, when Cassini married the 30-year-old noble lady Geneviève de Laistre, "*moglie secondo il mio cuore*" or "wife after my own heart", as he wrote some years later.

Hers father was the lieutenant general of the comté of Clermont in Beauvaisis and a King's advisor, and Colbert and the King himself attended the wedding as witnesses. The astronomer, as we will see in a moment, was already a French citizen and became also owner of the Castle of Fillerval (probably from the assets of the mother-in-law) in the county of Thury-sous-Clermont, and the presence of the King at his marriage, according to Christiane Demeulenaere-Douyère in his work "*La famille Cassini et l'Académie des Sciences*" [The Cassini family and the Academy of Science], means that: "... *Cassini devient aussi un Courtisan*" [... Cassini becomes also a courtier].

Cassini Dynasty

When he got married, Cassini was a French citizen, because in April, some months before the wedding, the King had discussed with him the opportunity to reside in Paris or any other city in France, enjoying the rights of French citizenship by birth. Evidently, the proposal was promptly accepted since Cassini's "*Lettres de naturalité*", or letter of naturalization, was filed on June 14, 1673, signed by Louis XIV. Becoming a French citizen, the scientist, now Jean Dominique Cassini, ceased any legal link with Italy (Fig. 17.1).

Cassini and his wife Geneviève had two sons: Jean-Baptiste and Jacques. Both were taught mathematics, but Jean-Baptiste would be killed in 1692, in the battle of La Hauge, while Jacques would become an astronomer known also as "Cassini II". This is because, as reported in "*Observatoire de Paris, son histore 1667–1963*" (Observatoire de Paris, 1984), the Cassini "reigned" in the Observatory as a "*Véritable dynastie, puisque tells des souvrains les nom de chacun d'eux est suivi d'un numéro, les Cassini ont contribué par leurs travaux et leurs découvertes au progress et au renom de la science astronomique et géodesique française*" [true dynasty, since as is the case for the kings, the names of each of them are followed by a number, the Cassini contributed by their works and their discoveries to the progress and renown of French astronomical and geodetic science].

We will return to his descendants later, limiting our account for the moment to mention that Jean-Baptiste was born in 1674 and Jacques on February 16, 1677, both in the Observatory. The father, beginning in their childhoods, directed both his children to astronomical observations and the study of geometry through the *Elements*, the fundamental book of Euclid. The documents regarding the theses in mathematics of the two sons are preserved in the Library of the Paris Observatory.

Fig. 17.1 Statue of Jean Dominique Cassini at the Astronomical Observatory of Paris (Author's photo)

The subject was optics, and they discussed their work first on August 10, 1691, at the Mazarin College, and later at the Observatory on September 2 of the same year.

Actually, only Jacques followed his father's footsteps; Jean-Baptiste instead pursued a military career and died young, on May 29, 1692, in the naval battle near

Le Hauge between French and English forces, when he was on board a French ship with the rank of midshipman. In the Cassini journal he reports it, amid observational data, with the simple words: "*today my son Jean-Baptiste died*".

Director or Not Director

It is often written that Cassini became the director of the new astronomical observatory of the Academy of Science in Paris. However this is just some "common knowledge", for this position did not actually exist in his time. When Colbert built it, he did not appoint a director in charge or, as noted in the writings of his great grandchild Cassini, namely Cassini IV, allocate any funds for the maintenance of the building or of the astronomical instruments.

There were also no funds to pay people who dedicated themselves to astronomy, with the exception of the only officially recognized and remunerated, the so-called "*concierge*", a sort of guardian and conservator. Such an assignment was first entrusted to Couplet, who became the first French assistant of Cassini, and after him to the nephew of the astronomer himself, Giacomo Maraldi.

Even though Cassini was not entitled of a salary as a director of the observatory, or for any other staff position in this institution, he enjoyed a Royal grant of 9000 "livres" (the currency used at the time). The significance of this sum can be better understood by comparing it with the salary of 6000 livres remitted to his colleague Huygens as president of the Academy of Sciences, with those between 1200 and 2000 livres of the Academy members, and with the 600–1000 livres for the "pupils" of this institution. It is possible that Louis XIV strategically intended to impress the European countries with such a remarkable sum, and apparently he reached his goal if it is cited accurately in a letter by Henry Justel to Henry Oldenburg, secretary of the Royal Society of London and creator of the scientific peer review: "... *On luy donnera neuf mil livres par an et il sera logé*" [He (Cassini) will be given nine thousand livres a year, and he will be lodged].

Indeed, this situation of an Observatory without a director can be understood by reading these lines of Charles-Joseph-Etienne Wolf in his "*Histoire de l'Observatoire de Paris, da sa foundation à 1793*" (Paris, 1902): "*Une Académie dont tuos les membres ont des droits égaux est impuissante à diriger un Observatoire*", an Academy whose members have all equal rights is helpless to manage an Observatory, since their members "...*travaillant selon sa convenance et se goût, suivant l'inspiration du moment*" work according to their convenience and their taste, following the inspiration of the moment. In other words, the observatory had no director because a director needs a staff to be directed, something that the Academy and their totally independent members could not accommodate at all.

While this situation may look odd to our modern eyes, at those times it was just the norm. One has to remember that the Academy was based on the model of the free association of scientists that were starting to rise in Italy and in Europe in those years, and that the Paris observatory had not been conceived as a dedicated institution, but rather as a multi-purpose building capable of hosting the meetings of the Academy members, as well as their experimental facilities.

So the Academy had no precise line of research to follow, and we owe it to Cassini, to its modern view of organized research work, and to its role of director de facto if the Observatory started with a well-established observation plan that ensured its functioning. The importance of his role can be understood by mentioning how the building, after him and Colbert, began to decay because of the lack of the necessary maintenance. This process had advanced so much that, in 1771, Cassini de Thury or Cassini III officially appointed director of the structure, immediately decried the existing collapses and water infiltrations.

Curiosity

Despite her name, Anna Cassini is not related to the famous astronomer, as his descendants became extinct beginning in the 19th century. She, however, is the author of "*Gio: Domenico Cassini*", and of other books in Italian, in which she presents a rich collection of archive documents about him and the Maraldi family. Thanks to her careful archival researches, we know that the letter of naturalization mentioned in this chapter was filed on June 14, and not in July as reported by other sources.

Another correction to what can be found in various biographical sources is about the year of the marriage, which was in 1673 and not in 1674. This is what can be found at Clermont, in the Archives of the *Société Archéologique et Historique de Clermont-en-Beauvaisis*, where at pages 95–96 the "*Document notarié de la généalogie de La Myre-Mory*" reads: "*...en l'année 1673 à Paris, il épousa Demoiselle Geneviève de Laistre...*" [in the year 1673 in Paris, he married Miss Geneviève de Laistre].

Jean Dominique Cassini and Geneviève lived in the large apartment on the first floor of the Observatory of Paris, and there were born their children, those of their children, and other descendants for four astronomical generations until the French Revolution. Actually they had three children: "*...une fille et deux graçons...*" [a girl and two boys] as reported in handwritten notes by Cassini IV sent to the canon Fabroni for his work "*Vitae Italorum doctrina excellentium...*".

It seems that the King Louis XIV with his court habitually visited the Observatory at night, and that Mrs. Cassini had to wake up all of a sudden and come down from her apartment on the first floor to receive the sovereign and other guests.

Anne-Tullie Cassini, the only daughter of the couple, was born June 2, 1678. She was baptized with the names of her grandmothers in the church of Saint Jacques du Haut-Pas, in the Observatory parish. Unfortunately she died at an early age, by no means an unusual event at the time.

The castle of Fillerval in Thury was the country residence of the family, but Cassini did not stay long there because of his work in Paris. After the death of the father, his son Jacques bought the castle and the surrounding lands, becoming the gentleman of Thury. He spent long periods in the castle and built an observatory in one of its towers, which had previously been used as a water reservoir.

The meridian of Paris passes very close to Thury, and today the castle, thoroughly restored, is owned by the *Institut Française de Gestion* (French Institute of Management).

Chapter 18
Cassini Maps

On travailloit alors dans l'Académie
à la mesure d'un degré
de la circonference de la Terre...
Giovanni Domenico Cassini

Carte de Cassini

One of the main reasons for which Cassini was called to Paris, as previously mentioned, is related to cartography. The Academic Jean Picard, who was carrying out some geodetic surveys for the new "*Carte de France*" (Map of France) commissioned by Colbert, realized the importance of the work of the Italian astronomer for the measurement of longitudes.

The breakthrough was his fundamental work *Ephemerides bonomienses mediceorum syderum*, published in Bologna in 1668, which was read not only by the astronomer and geodesist Picard but in all of Europe. It reported the exact times of the eclipses of Jupiter' satellites, calculated for the meridian of Bologna. These particular astronomical phenomena were so accurately observed by Cassini that these bodies could be used like a celestial clock that made it possible to calculate the difference of the meridians from one place to another; in other words, his method could be used to determine the exact longitude of any site.

The importance of realizing the first topographical map of France was fundamental for both economic and political matters and obviously also for the military. The work was very long and could be completed only by Cassini's descendants in 1789 (or even in the early 19th century, if one considers the printing dates of the last maps of the final atlas) under the name *Carte de Cassini* [Cassini Map] (Fig. 18.1).

So Giovanni Cassini, as soon as he arrived in Paris, immediately started to help Picard in his mission, as evidenced by the minutes of the Academy, in which his name appears for the first time on July 31, 1669: "*M. Picard [...] a lu un mémoire contenant la relation de son voyage en ces termes: Ensuite de la resolution prise à l'assemblée d'aller à Mareuil pour y verifier la position des principaux points qui doivent servir comme de fondaments à la Carte des environs de Paris, nous nous y*

G. Bernardi, *Giovanni Domenico Cassini*, Springer Biographies,
DOI 10.1007/978-3-319-63468-5_18

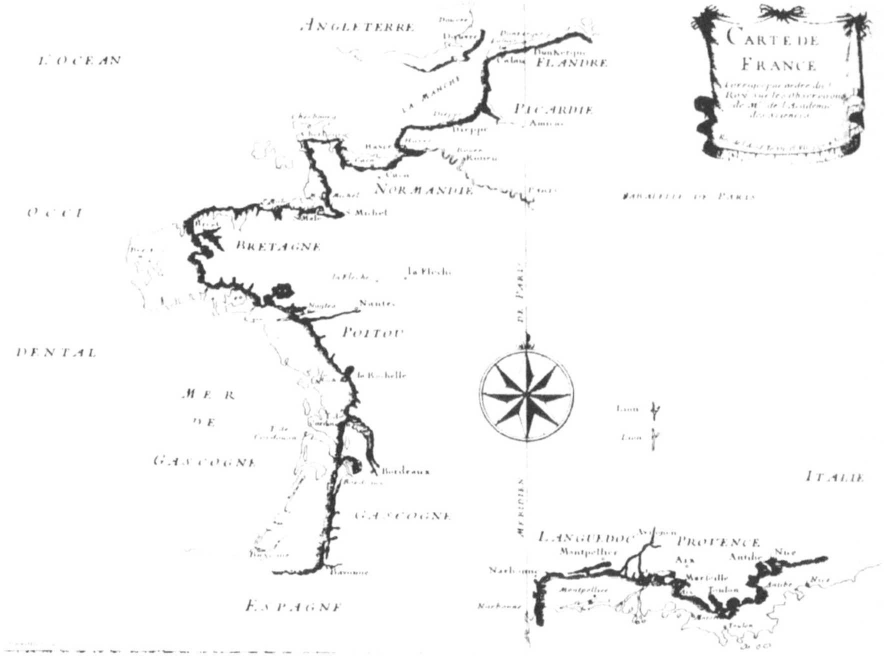

Fig. 18.1 First map of the coasts of France based on astronomical surveys, established by Picard, Lahire, and Cassini (1682)

sommes le 26 juillet, Mrs. Cassini, Richer et moi, transportés et nous avons trouvé le lieu autant commode à notre dessein qu'on le pouvait souhaiter et, quoique le temps ne fût pas fort favorable, nous n'avons pas laisse de prendre au juste les angles de positions de divers lieux à la ronde" [M. Picard read a dissertation containing the report of his travel in these terms: after the decision taken at the meeting to go to Mareuil to verify the position of the principal points that are to serve as foundations for the map of the environs of Paris, we were there on the 26th of July, Mrs. Cassini, Richer and myself, and we found the place as convenient to our design as we could wish, and, although the weather was not very favorable, we have not omitted to record precisely the angles of positions of various surrounding places].

Cassini remembers as well in his memoir this first expedition with Picard, but he also describes how the surveys were made: "*On travailloit alors dans l'Académie à la mesure d'un degré de la circonference de la Terre [...] nous allames avec Mr. Picard a passer quelques jours a Mareuil [...] et nous determinasme pendant la nuit la situation de la ligne tirée de Mareuil à Clermont par reule moyen de la plus grande declinaison de l'etoile polaire par la distance de cet etoile au pole et par le hauteur mesme du pole...*" [we worked in the Academy to the measure the degree of the circumference of the Earth ... we went with Mr. Picard to spend a few days at Mareuil ... and we determined during the night the situation of the line drawn from

Mareuil to Clermont by means of the greatest declination of the polar star by the distance from that star to the pole and by the exact height of the pole].

Cassini began this work in 1670, using the triangulation method that had been introduced for the first time in the modern era by Tycho Brahe, to create the topographic map of France or *Cassini Map*: a composite map made of 182 sheets that could be joined together to form a physical map about 39-ft high by 38-ft wide. This impressive atlas, however, was the result of the very long work comprising the first national cartographic survey, and which involved four generations of the Cassini family.

In 1681, Picard, with the collaboration of Jean Dominique Cassini, completed the measures of the French meridian and presented it to the Academy. This was a very ambitious project to extend his initial achievement: "*Project de la prolongation de la méridienne jusqu'aux deux mers pour la mesure de la Terre*" [Project of the prolongation of the meridian to the two seas for the measurement of the Earth] or in other words the map of the entire France, the future Cassini Map. Unfortunately, Picard died the following year. Later in the same year, the "Corrected Chart of the France" (traced as a point of reference for the meridian of Paris) that can be considered a prototype for the second project was presented to the Academy. So Cassini took on himself the task to finish the project initiated by Picard, namely the measures of the meridian passing through Paris. By his intention, the measurements to the north as far as Dunkerque would have been done by de La Hire, while those to the south as far as Perpignan by Cassini. The death of the major sponsor of this project, the Prime Minister Colbert, stopped the project once again. It was resumed in 1700 when the new Prime Minister started anew to support the idea, and this work was brought to completion by Cassini II in 1718.

The complete survey of the country, however, would be carried out only between 1750 and 1789 under the direction of Cassini III and Cassini IV, and the 182 sheets at the same scale (1:86,400) of the map were published as an atlas between 1756 and 1815. This was the first modern map with accurate longitude and latitude estimations based on the latest scientific achievements of the time.

You can see the whole 182 pages of the atlas of the Cassini map laid over Google Maps on the David Rumsey Historical Map Collection (http://rumsey.geogarage.com/maps/cassinige.html). The Map does not precisely locate the current buildings or the boundaries of the forests, but the level of precision of the road network is remarkable.

The Problem of the Longitude

One of the priorities of Luis XIV was to make France the most powerful state in the world, a purpose encouraged by the minister Colbert who exploited the skills and the knowledge of Cassini to this end. As part of this goal, Colbert had specific funds for missions of various astronomers to calculate the longitudes of various places abroad, another project that lasted for several years and started with the arrival of Cassini.

Having precise maps, and knowing how to determine precisely your position on the Earth, but especially at sea, was not just a matter of avoiding becoming lost. Rather, it meant the control of commercial routes, avoidance of disastrous shipwrecks, so frequent at the time, and efficient movement of your armies and fleets. In short, the main consequences for a nation encompass the commercial, economic, and political spheres.

As mentioned earlier, determining the latitude of a place is relatively easy; the real problem was rather that of the longitude. The method used to calculate the longitude was based on Cassini's theory, suggested by Galileo studies on the eclipses of the moons of Jupiter, as the now French astronomer clearly explains: "*… les phases le plus propres pour determiner les longitudes géographiques sont les immersions des satellites dans l'ombre de Jupiter et leur émersions de l'ombre…*" [the most suitable phases for determining geographical longitudes are the immersions of satellites into the shadow of Jupiter (namely the occultations, A/N) and their emersions from the shadow].

To this end, several expeditions departed from the Academy of Science and Paris, under the personal coordination of Cassini. He also collected and organized the results of the astronomers' measurements. Indeed, among other qualities, this "non-director" found himself to be an excellent work manager. Cassini also improved the scientific instrumentation provided to the astronomers, especially to make them more reliable and prevent any breakdown; moreover, guided by his unprecedented experience, he wrote a booklet with detailed instructions about field operations.

The importance of the "*Instructions générales pour les observations géographiques à faire dans les voyages*" [General instructions for geographical observations to be made during travel], cannot be underestimated. With this memo, provided to the expeditions, Cassini ensured that the astronomers traveling any place on the Earth could make the measurements of longitudes following what we would now call a "standard procedure", which ensures the most accurate and homogeneous results. Today, having such documentation is deemed essential to coordinate the work of a scientific project, but, at the time of Cassini, the concept of "*observer de concert*", or observing together, was completely new and exceptional.

The first of these expeditions had to be to Egypt, a symbolic target for Cassini who, in the land of the Pharaohs, wanted to pay tribute to the famous astronomer Ptolemy of Alexandria. However, more practical reasons prevailed instead, and the Academy started from the isle of Hveen or Ven (Fig. 18.2). In this place, today a Swedish island, another famous astronomer, the Dane Tycho Brahe, had established the famous Uraniborg, the first European astronomical observatory and organized his systematic observations there.

The motivation was clearly explained by the Academic and priest Jean-Felix Picard (1620–1682) years before, in a session of the Academy: "*…pour…substituer le meridian de Paris au lieu et place de celuy de Uraniborg, il est necessaire de sçavoir exactement la difference de longitude, qu'il y a entre ces deux meridiens, et* *pour cet effect il faudroit avoir des observations des satellites de Jupiter correspondemment faictes dans ces deux lieux*" [in order to substitute the meridian of Paris instead of that of Uraniborg, it is necessary to know exactly the difference in longitude between these two meridians, and for this effect it would be necessary to have observations of the satellites of Jupiter correspondingly made in these two places.]. In other words, if Paris had to become the "origin of the longitudes", it was necessary to know exactly the longitude of this island with respect to Paris, because the most complete and precise catalogue of the time referred to this location.

Picard is considered one of the founders of modern astronomy in France, and despite his age on July 21, 1671, left for Hveen Island to Uraniborg, firmly convinced by the scientific rationale cited above.

So, especially for Picard, Cassini compiled what he defined as a "*précieuse list*" [precious list] of the eclipses of the satellites of Jupiter which he would have observed from Tycho's island until the end of the same year. Cassini, at that time, had been in Paris for a couple of years, and, as reported by Kurt Pedersen in his "*Une mission astronomique de Jean Picard: le voyage d'Uraniborg*": "*A cette époque Cassini était à peu près le seul astronome à pouvoir calculer à l'avance les éclipses des satellites de Jupiter avec une precision suffisante*" [at that time Cassini

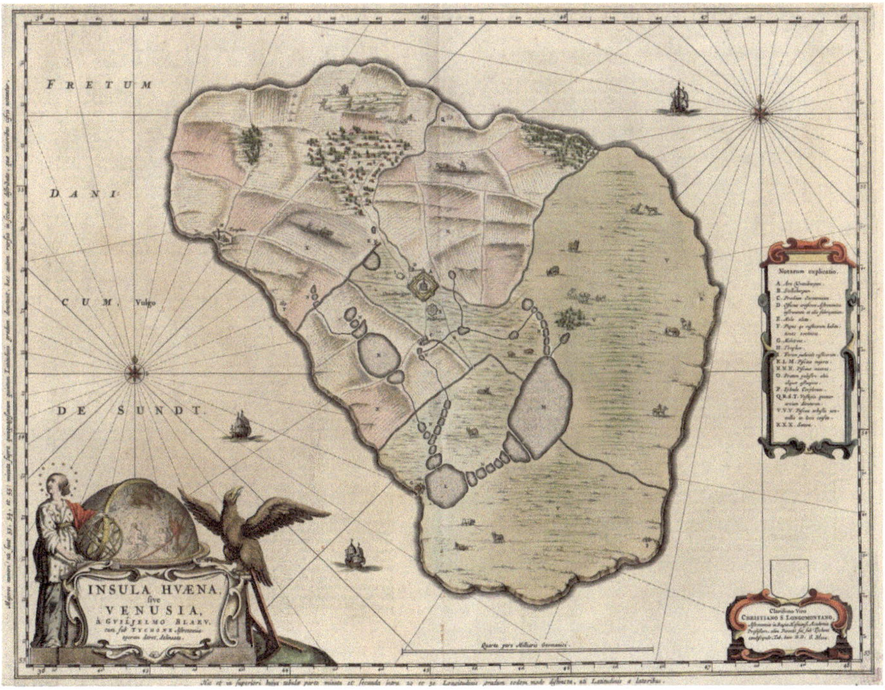

Fig. 18.2 Map of the Island of Hven from a copper etching of Willem Janszoon Blaeu's (1663). The Uraniborg observatory can be seen just above the center

was almost the only astronomer able to calculate in advance the eclipses of the satellites of Jupiter with sufficient precision].

Unfortunately, Picard stay on Hven, deserted by colleagues, deprived of communications, and beaten by the winds, was not comfortable at all. Cassini did his best to support the morale of his colleague, who also became ill, and the situation would improve thanks to the improved methods disseminated by Cassini, which made possible for the first time the remarkable success of these observations. When Picard had finished, on Cassini's suggestion, he went to Danzig, on the return trip, to see the astronomical instruments of Johannes Hevelius and his famous private observatory.

After the success of this first scientific mission, in the following years other astronomers left for far away countries. For example, the Academic Jean Richer (who began the science of gravimetry) went to Cayenne in Guyana in 1672, and this too was a great success. Here, near the Equator, he could also use the opportunity to study other phenomena, like atmospheric refraction and the calculation of the parallax of Mars. In particular, Richer in Guyana, Cassini in Paris, and in England John Flamsteed, the first Astronomer Royal, observed Mars simultaneously during an opposition, namely when Mars, the Earth, and the Sun are aligned in this order, taking advantage of the large distances of the observers to determine the parallax of the planet.

These data were used to calculate the distance between Earth and Mars, and to determine the value of the Astronomical Unit (AU about 150 million km) or the distance between the Earth and the Sun. Flamsteed calculated a value of 131-million km and Cassini 140-million km, a historical moment when planetary distances would be greatly improved with respect to past estimations.

Chart de la Lune

On February 18, 1679, Cassini presented to the Academy of Paris his "*Chart de la Lune*", that is, the Map of the Moon, which was the first scientific map of our satellite. Engraved on a copper plate the following year by Jean Patigny, it is the result of an observational research program which lasted for nine years. It started on September 14, 1671, with an objective made, as usual, by Giuseppe Campani, having a diameter between 8 and 10 cm and 5.5 m in focal length. The next year it was continued with more accuracy with the new Campani telescope, with a twice as long focal length of 34 ft. During these years Cassini, with the collaboration of his artistic assistants Jean Patigny and Sebastien Leclerc, produced some 60 drawings of the lunar surface that were used as the basis for the Map of the Moon.

This masterpiece remained unmatched in accuracy and precision until the 19th century, with the introduction of photographic plates. The representation of the Moon has a diameter of about 21 "*pouces*" [inches] (54 cm); it is illustrated with unprecedented accuracy of the topographical details, and the quality of the representation of the light and the shadows is so high that it creates a three-dimensional

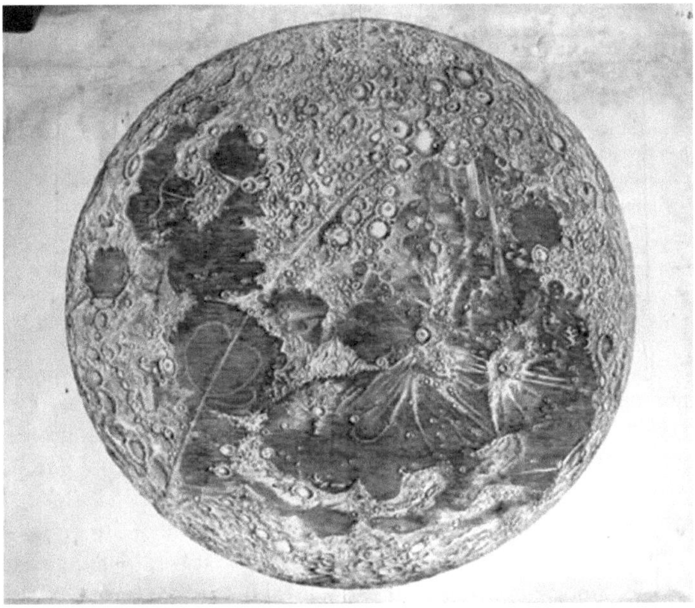

Fig. 18.3 Map of the Moon by Cassini (Observatory of Paris)

effect. The astronomer made the first original drawings based on direct observations at the Observatory of Paris; beginning with his arrival in 1671 and over the next eight years, he systematically studied our satellite, as witnessed by the date, time, and all the observational data reported with each drawing (Fig. 18.3).

These valuable drawings are now in the Library of the Paris Observatory, but they were bound by his grandnephew Jean Dominique Cassini IV into a volume of more than 60 pages "*…parfaitement dessinées de la main de le Clerc…*" [perfectly drawn by the hand of Leclerc], but, in addition to these beautiful tables, the French astronomer carried out a very detailed study of the lunar motions, in order to determine with better accuracy the inclination of the lunar equator and the orientation of its orbit. In 1693, Cassini also formulated a "*Nouvelle Théorie de la Lune*" [New Theory of the Moon] also known as the "Three Laws of the Moon", in which for the first time the motions are described in detail with respect to the ecliptic. It was so accurate that, more than one century later, the astronomer Pierre Simone Laplace, famous for his *Celestial Mechanics*, still mentioned it as one of the finest studies by Cassini.

In particular, the three laws, as reported in the "*Traité de mécanique celeste*" by Félix Tisserand (1891), state:

1. *La Lune tourne sur elle-même, dans le sens direct, d'un mouvement uniforme autour d'un axe dont les pôles sont fixes à sa surface; la durée de la rotation, 27j 7 h 43 m 11,5 s est identique à la révolution sidérale de la Lune autour de la Terre.*

In modern language, we now say that the Moon has a 1:1 spin–orbit resonance. This implies the well-known fact that the orbital and rotational motions of the Moon are such that, on average, our satellite shows always the same side to the Earth.

2. *L'axe de rotation fait un angle constant avec l'écliptique; cet angle est de 88° 25′*
3. *L'axe de l'écliptique, l'axe de l'orbite de la Lune et son axe de rotation sont constamment dans un même plan*

The second and the third laws, in contrast, are nothing else than the discovery of a phenomenon now called *latitude libration*, namely, that the Moon rolls slightly back and forth in the latitudinal direction, thus showing a little more than half of its surface to the Earth.

A reduced version of the map was published for the lunar eclipse of the July 28, 1692, because Cassini, in issue of June 30, 1692, of the Academy Journal, had alerted the scientific community about the importance of this event: "*...car elle se fera sur l'horizon occidental dans une partie de l'Europe; de sorte que l'on pourra voir en même temps sur l'horizon la Lune éclipsée et le Soleil...*" [for it will be on the western horizon in a part of Europe; so that we can see at the same time on the horizon the eclipsed Moon and the Sun]. In practice this is made possible by the refraction of the atmosphere. The Sun and the Moon, in fact, are separated by 180° during a lunar eclipse, but, when both the objects are close to the horizon, their separation is apparently less because of the refraction. Thanks to this favorable conditions, and with the help of his map, astronomers could make accurate observations about the times when specific spots listed in the paper and marked on the map entered into or exited out of the shadow of the eclipse. As pointed out by the French astronomer, this would have helped to improve the estimation of many features that influence the evolution of an eclipse, like the apparent diameters of the Sun and of the Moon and their parallaxes. Finally, both the reduced map and his observation data about the eclipse were published in the Memoirs of the Academy.

Curiosity

Only in 1700 did Cassini and his family, including his wife, son and nephew, travel to the Pyrenees to resume the measurements of the project of the meridian line of Paris, and, in Bourges, a pillar was erected indicating the exact position of the meridian in that city. In 1701, the expeditions finished their measurements, Cassini is 76 years old, and this will be his last working journey.

The toponyms called "*Signal de Cassini*" [Signal of Cassini] are points of reference corresponding to the vertices of the large number of triangles that form the reference points of the *Cassini Map* (Fig. 18.4). Today, you can see them marked on the ground, revealing the places where the measurements were carried out at that time.

Cassini III founded in 1756 a company with fifty members belonging to the aristocracy and the high officers of the nation to raise the funds needed to complete the topographic surveys. The most famous among them was the Marquise Pompadour, the official chief mistress of Louis XV and the most powerful woman of France.

Napoleon Bonaparte in 1808 decided to make a new Map of France, but the project could not get started. It began only in 1817, lasting until 1866, and the result was the *Map of the Etat-Major*.

Jean Picard published in 1671 his "*Mesure de la Terre*", and he may have been a potential inspiration for the fictitious Star Trek character with a similar name (https://www.seeker.com/star-trek-inspiration-meet-the-real-jean-picard-1765425621.html).

The Academy financed others scientific missions in distant countries like Senegal, Martinique, the Cape Verde Islands, and Egypt. In other cases, like for Thailand or China, Cassini was helped by Jesuits missionaries because they had very good mathematical and astronomical training, which they also used as a help in their mission to convert the population.

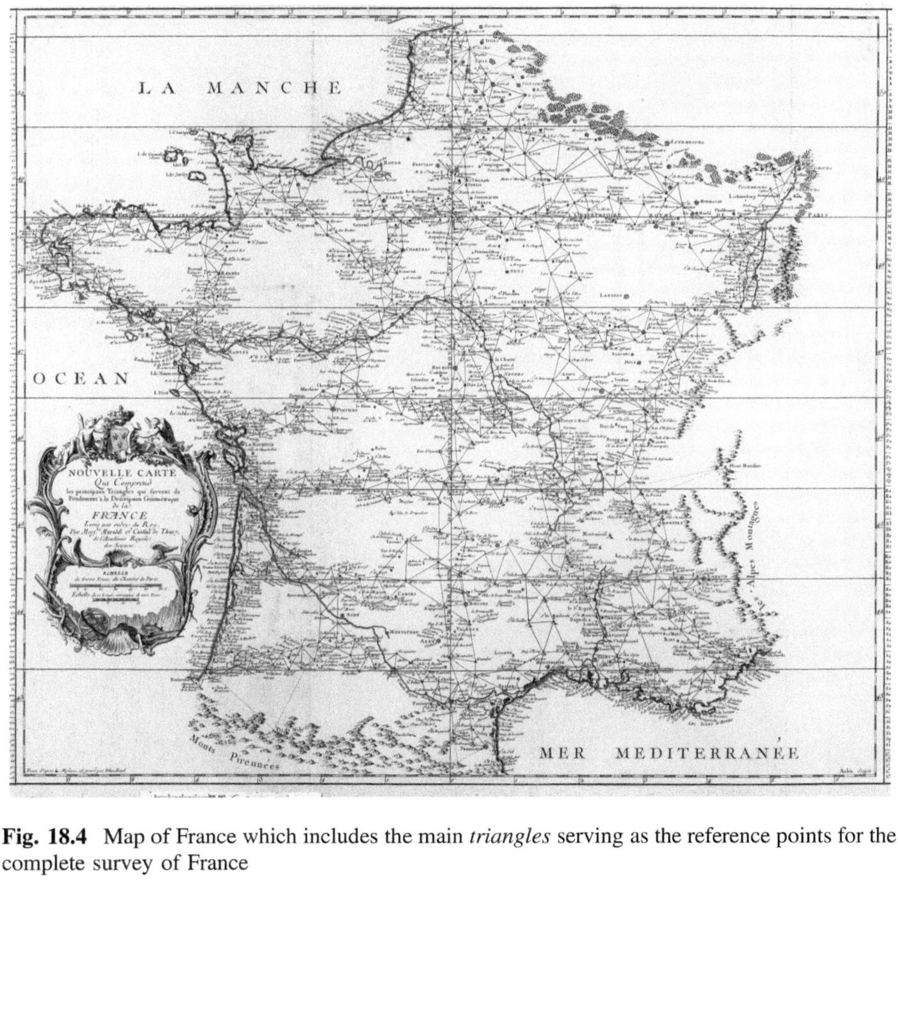

Fig. 18.4 Map of France which includes the main *triangles* serving as the reference points for the complete survey of France

Looking in greater detail at the scientific-artistic masterpiece of Cassini's Moon, in the bottom right corner near Cape Heraclides on the Sinus Iridum aka Bay of Rainbows, you may notice *la tête de femme de la Carte de la Lune de Cassini*, that is, what looks like the head of a woman known also as "*the Moon Maiden*". This is commonly thought to be a portrait of his wife Geneviève Delaistre, who he had married in November 1673. As reported John Lockerbie: "*Confirmation of this is tenuous, however, in 1678, Jean Dominique commissioned a pen-and-ink portrait of his wife from Jean Baptiste Patigny, the son of the artist and engraver of Jean Dominique's map of the Moon of 1679, which gives some credence to the possibility*."

On the floor of the west tower of the Observatory was a huge map of the world and on May 21, 1682, the King with the Queen, the Dauphin, and the whole court paid a visit to Cassini and other scientists to be updated on the progress of the research, but they especially admired, in particular, this map. It had been drawn, under the supervision of Cassini, by Sedileau and Chazelles, two of his pupils, with a washable ink, useful for any corrections or additions, and it does not exist anymore.

Chapter 19
Journey to Italy

> *...nous ne nous sommes pas contenté dans*
> *ce voyage de determiner la longitude*
> *et la latitude des lieux*
> *où nous avons fait quelque séjour...*
> Jean-Dominique Cassini, "*Observations astronomiques faites dans les voyages de France et d'Italie*", Mémoires de l'Académie des Sciences, t. VII, Paris, 1729

In the early autumn of 1694, Cassini, close to his seventies, set off on a long journey to Italy with his young son Jacques in order to determine the geographical coordinates of several towns in the peninsula and to restore in Bologna the meridian line in the church of San Petronio. He left behind for a few years in the Paris Observatory his nephew Giacomo Filippo Maraldi, later known as Jacques Phillipe Maraldi or Maraldi I.

He was the son of the astronomer's younger sister, and, in 1687, he had been invited by his famous uncle to join him in Paris to complete his mathematical and astronomical studies. Eventually, he became a valuable assistant astronomer, as reported in the chapter about the Maraldi branch of the "Cassini dynasty", and in 1694 he was able to make regular and careful observations in the Observatory. His ability is testified to, for example, by an excerpt from a letter sent by his uncle from Rome: "*Les taches que vous avez observes dans le Soleil non obstant leur petitesse font connaitre votre attention et assiduité aux observations...*". Here Cassini congratulates him on recent observations of some small sunspots, stressing how this reveals the attention of his nephew to the observation task.

The French astronomer and his son left Paris on September 23, 1694, in a carriage loaded with scientific instruments like an octant, a number of telescopes, and a pendulum, making a first stop-over at Fontainebleau, to pay homage to the King who was there with his court. Here Cassini received the permit to cross the borders to Italy, travelling from Burgundy to Provence, passing through the Rhône valley. In southern France, they passed through Aix-en-Provence, Marseilles, Toulon, and finally reached Nice, where Cassini made some observations in the

G. Bernardi, *Giovanni Domenico Cassini*, Springer Biographies,
DOI 10.1007/978-3-319-63468-5_19

governor's palace. Here, Cassini determined the geographical coordinates of the place, comparing his data with those calculated simultaneously by Maraldi in Paris.

The Birthplace

Cassini and his son, in early November, crossed the border and stopped in Perinaldo, the astronomer's birthplace. This stop was not just for a visit to the Italian relatives; rather, Cassini had also a scientific purpose: here in fact, from the Ligurian hills, he wanted to test a new approach for the determination of distances. Unfortunately, the system of triangulations introduces errors due to atmospheric refraction, but Cassini and his son: "*...nous ne nous sommes pas contenté dans ce voyage de determiner la longitude et la latitude des lieux où nous avons fait quelque séjour, mais nous avons travaillé aussi à faciliter les descriptions particulières qui se font par la Géometrie pratique, en nous servant de quelque méthode qui n'avoit point encore été pratiquée*".

He recounts that they have not been satisfied in this journey to determine the longitude and latitude of the places they had visited, but they have also worked "to facilitate the specific descriptions that are made in practical Geometry", and that they "used some methods which had not been tested yet." He was referring to the effect on geodesic measurements of the refraction of light, namely the deviations suffered by light rays when they pass through different media. As discussed in the chapter about the meridian line of San Petronio, he had already greatly contributed to the understanding of this phenomenon, but now he was seeking a fixed ratio useful as a "*...correction des rayons visuels...*", i.e., a correction factor.

The hills of Perinaldo, with their unobstructed horizon towards the sea, are ideal for observations aiming at "*...verifier ces mesures par une méthode particulière*", [verifying these measures by a particular method], and the object chosen as the target for the measurements is called, in the astronomer's logbook, a "Bastion". It is a type of tower built in the mid-16th century for spotting pirates or Turks and, according to Anna Cassini, the one used by Cassini is probably the Alpicella's Tower, whose altitude and distance are in close agreement with those from the coast provided by Cassini.

The journey continued, and, after Perinaldo, Cassini and his son stopped in Sanremo, Savona, Genoa, Portofino, Livorno, Pisa, and Florence. Here the two astronomers, with the collaboration of Vincenzo Viviani, an old friend of Giovanni Cassini mentioned previously, on December 16, near the Florence Cathedral of Santa Maria del Fiore (in English "Saint Mary of the Flower") with the proven method of the times of eclipse of the first satellite of Jupiter, established the longitudinal difference from Paris.

The two also left Tuscany, reaching Bologna through the pass "Raticosa e Pietramala" a difficult road through the Apennines. Known before the Romans, the pass was the only road connecting Florence and Bologna. It had to be crossed with horses or mules, and the trip required two days, but, despite his advanced age, for

Cassini this was not a problem. In the past, he has become accustomed to travel in every season along this difficult path because of his past services in Bologna, Rome, and Florence, a training that evidently he remembered very well.

The Time and the Spiritual Power

Just arrived in Bologna, Cassini went to San Petronio to check the status of the meridian line. Here he could examine his instrument after 40 years; the floor near the columns had lowered, probably because of some earthquakes, but, despite this inconvenience: "*La Meridiana si è conservata nella primitiva situazione...*" the meridian line has been preserved in the original condition! With this sentence, the astronomer meant that it had not been damaged by the structural changes of the church.

The instrument, however, required a restoration to return it to its previous accuracy, so Cassini began a leveling work to reposition the iron line, but "*Le neiges qui continuent encore icy me font perdre les plus belles occasions d'observer...*", [The snow that is still falling here makes me lose the best opportunities to observe...]. Actually, in January 1695, Bologna experienced some heavy snowfalls, and he lost the possibility to do all the scheduled observations, but he was a creative.

He had ordered the construction of "*...une espèce de pinnule...*" that is a sort of target, mounted on a graduated slider, and this, used together with the meridian line, enabled him to determine the height of the pole, or the latitude of the place, observing the North Star inside San Petronio. This particular instrument for the observation of the polar star, which Cassini simply called "macchina" or machine, had been designed by the astronomer, but it was built and delivered by a craftsman when he was in Rome. Today, it is preserved in the Museum of San Petronio.

The visit of the two Cassinis in Bologna was longer than expected because of the snow, but in spring they were in Rome again, to make other astronomical surveys. In addition to this, Cassini met the Pope and participated in a conference with Cardinals and experts for the old "water problem" about the rivers Reno and Po, thoroughly discussed in a previous chapter. In fact, this had remained unsolved for all this period, and Cassini, unsuccessfully, tried to push through the only possible solution: follow the natural course of the rivers!

In the autumn of the same year, it was time to return to Paris, and Cassini retraced his journey, stopping shortly in Bologna just long enough to check the meridian line and say goodbye for the last time. In Genoa, he met his fellow student Francesco Maria Imperiali-Leorcaro, now the city "Doge", that is, the most important political authority of the republic, and "*...Le 6 Décembre ... nous allâmes le matin avant le lever du Soleil, au lieu destiné pour faire l'Observation...*". Together, on the 6th of December in the morning before the rising of the Sun, we went to the place chosen for the observation, Villa Carbonara.

Here, the Prince Doria and other authorities took the opportunity to greet the scientist, and consulted Cassini for some problems about the harbor fort and the construction of a large meridian line in the Church of the Annunciation. This project, however did not go any further. Cassini continued his journey back and, in January 1696, he was in Perinaldo, to measure again the longitude of the place. As it appears from his report, he used the usual method of observing the eclipse of the first satellite by the shadow of Jupiter.

Who knows where he observed for the last time in his native country? From the terrace of his house of birth or in the village? We will never know, but his presence remained indelible among the inhabitants: an individual so important, a scientist at the French court, but most of all surrounded of special and mysterious instruments. Who knows for how many generations there was talk of his last visit in the village.

Curiosity

The fame of Cassini did not spare him from harsh criticism by colleagues. For example, Jean Baptiste Joseph Delambre (1749–1822), mathematician and astronomer, believed that Cassini's contribution to astronomy was over-rated. Nonetheless, he had words of admiration for the work of Cassini's nephew Maraldi I. Indeed, he condemned the behavior of the entourage at the observatory for not having honored him after his death: "*...on peut etre étonnée de l'indifférence que tous ses parens et ses confrères à l'Académie ont témoigné pour le travail qu'il laissait imparfait, et auquel il ne manquait plus que quelques étoiles circumpolaires*" [we may be astonished at the indifference that all his relations and his confrères at the Academy have shown to the work he had left imperfect, and of which only a few circumpolar stars were missing]. Here, Delambre refers to the almost finished catalogue of fixed stars to which Maraldi I had dedicated his life, as we will see in a later chapter.

The Church of the Visitation is a little church outside the village of Perinaldo, where tradition says that Cassini had given instructions for it to be built on the Ligurian Meridian. It is only a legend, since, when Cassini encountered the church, it had already been completed some time before, and it is not on the meridian. Moreover, no documents about the origin of this construction have been found yet.

Chapter 20
Cassini Dynasty

Giovanni Domenico Cassini, or Cassini I, thanks to his leadership, was the first of four successive Paris Observatory directors that bore this family name as a dynastic monarchy. They include his son, Jacques, or Cassini II (1677–1756); his grandson César-François, Cassini III (1714–1784), and his great grandson, Jean-Dominique, Cassini IV (1748–1845). Another descendant, the youngest son of Jean-Dominique, was Alexander-Henry-Gabriel or Cassini V (1781–1832), but even if he also studied astronomy, he devoted himself to botany. The birth and death dates are given in Devic, M.J.F., "*Histoire de la vie et des travaux scientifiques de J.D. Cassini IV*", Clermont 1851.

Cassini II

Jacques

We have previously mentioned the sons of Giovanni Domenico Cassini; the only one who survived to middle age, Jacques, was born in the Observatory on February 8, 1677, and was instructed in mathematics and geometry by his father beginning in childhood. Jacques then studied at the Mazarin College (in a previous chapter, we referred to the theses that he and his brother had defended) and he was admitted at the age of seventeen, as a student astronomer, to the Academy of Sciences.

In 1694, Jacques and his father set off on a long journey to Italy, and he participated in many expeditions in France to carry out the topographical surveys necessary for the drawing of the Map of France. He also travelled to England and Germany to complete his studies and was elected, in 1696, a fellow of the Royal Society of London and of the Academy of Berlin, becoming *maître des comptes* in 1706.

Jacques succeeded his father as director of the Observatory in 1712, and the following year he measured the arc of the meridian from Dunkirk in the north to Perpignan in the south, two towns at the extremities of France's latitudes, within 2°

G. Bernardi, *Giovanni Domenico Cassini*, Springer Biographies,
DOI 10.1007/978-3-319-63468-5_20

of longitude) and published in 1720 the results in a volume entitled *Traité de la grandeur et de la figure de la terre.*

He made two separate calculations for a degree of meridian arc, whose results were 57,097 toises de Paris (111.282 km) and 57,061 toises (111.211 km),

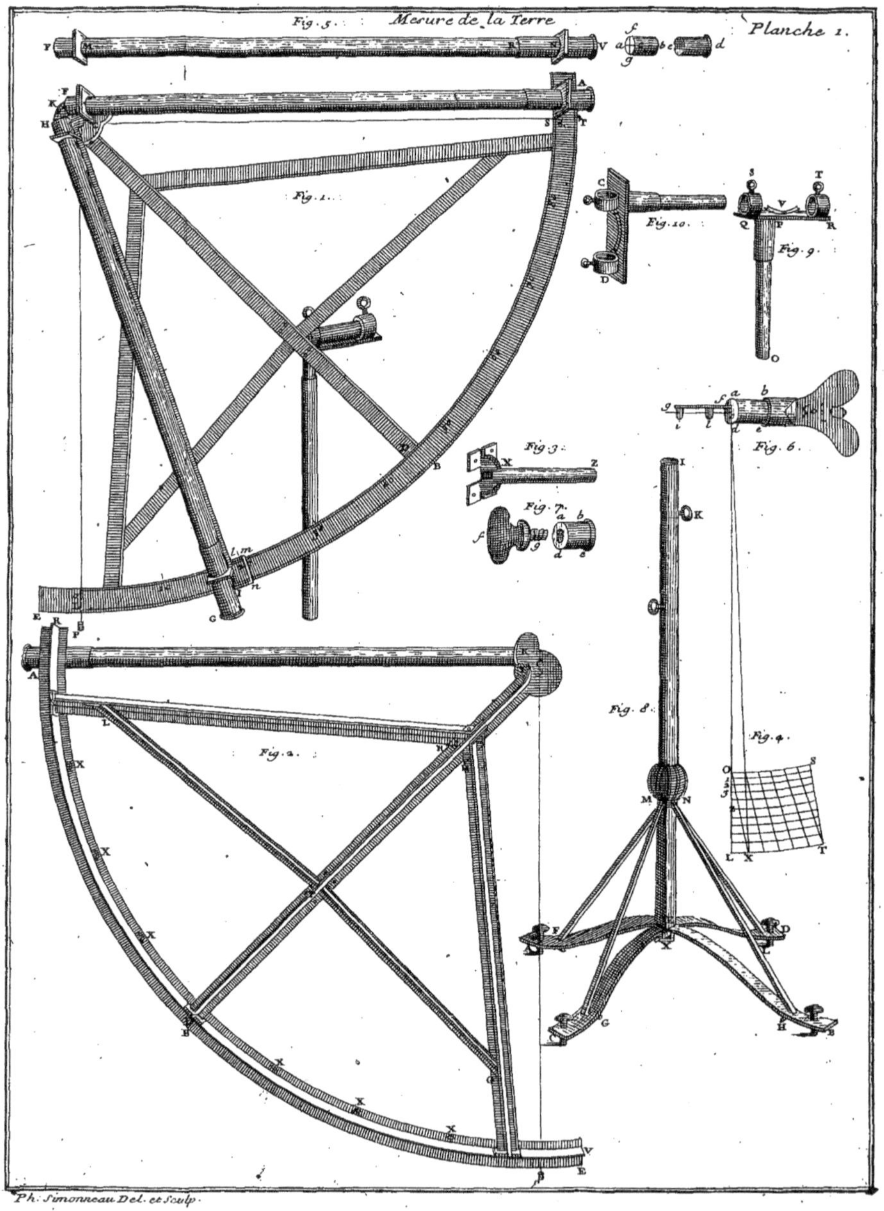

Fig. 20.1 Quarter-circle used by Jacques Cassini

resulting in Earth's radii of 3271,420 toises (6,375.998 km) and 3269,297 toises (6371.860 km), respectively (Fig. 20.1).

Cassini II also carried out research on Saturn, publishing the first tables of the ephemerides of its satellites 1716, and contributed to astronomical education with his book *Eléments d'astronomie*, published in 1740. He was also entrusted with important administrative responsibilities: *maître ordinaire de la Chambre des comptes*, then *magistrat à la chambre de justice et conseiller d'État.*

He married Suzanne Françoise Charpentier de Charmois, and, from 1740, gradually abandoned scientific activity, leaving it to their second son, César-François, also known as Cassini III. By contrast, the other son, the Marquis Dominique-Joseph de Cassini followed a career in the army. Elisabeth Géneviève Cassini and Suzanne Françoise Cassini were the younger daughters. Jacques died, aged 79, on April 18, 1756, in his mansion of Thury near Clermont, after a carriage accident. The asteroid 24,102 Jacquescassini is named after him, but unfortunately there are no portraits that depict him.

Cassini III

César-François

César-François was born on June 17, 1714, in Thury-sous-Clermont and studied astronomy, but he mainly devoted himself to topography and to geodesy problems, making accurate measures of the meridian of Paris. He became, in 1739, a member of the Academy of Science in Paris as a supernumerary adjunct astronomer. In 1741, he graduated to adjunct astronomer, and, in 1745, a full-member astronomer. In 1751 he was also elected a Fellow of the Royal Society in London (Fig. 20.2).

In 1747, the grandson of Giovanni Cassini married Charlotte-Jeanne Drouin de Vandeuil, daughter of Louis-François Drouyn de Vaudeuil, adviser of the king. They had two children: Jean-Dominique and Françoise Elisabeth.

Cassini III succeeded to his father's official position in 1756 and continued the hereditary surveying operations. Actually, with his research, he made a personal contribution to the drawing of the large *Map of France*, a great topographical map of France that represents a landmark in the history of cartography. This achievement would also be accomplished through his skills in overcoming the various political, economic, and practical obstacles to the project. The project would be completed by his son, Jean-Dominique or Cassini IV, and published by the Royal Academy from 1744 to 1793. Its 180 plates today are known as the Cassini Map.

In 1771, when the institution ceased to be a dependency of the Academy of Sciences, the post of director of the Paris Observatory was created for his benefit. The salary amounted to 3000 thousand livres, and César-François would obtain that

Fig. 20.2 Miniature portrait of French astronomer César-François Cassini or Cassini III [The Walters Art Museum (CC0)]

this right become hereditary. His chief works are: *La méridienne de l'Observatoire Royal de Paris* (1744), a correction of the Paris meridian; *Description géométrique de la terre* (1775); and *Description géométrique de la France* (1784), which was completed by his son Jean-Dominique.

César-François died on September 4, 1784, of smallpox in Paris.

Cassini IV

Jean-Dominique

Jean-Dominique, comte de Cassini, was born at the Observatory on June 30, 1748, and succeeded his father as director in 1784. He studied first at the Plessis College and later at the Oratorians in Juilly, sailing on the Atlantic Ocean in 1768 as "*commissaire pour l'épreuve des montres marines*" [commissioner for the test of marine watches] invented by Pierre Le Roy. He also travelled to Africa and America to make some surveys, which he later published in 1770.

In 1770, he was elected assistant astronomer at the Royal Academy, of which he became an associate member in 1785. The year before, he had been appointed director of the Observatory and devoted himself to its restoration and to the improvement of its instrumentation. His previously mentioned attempts at the restoration of the building, the reorganization of the place, and all his other initiatives were then ended by the French Revolution.

In 1783, he sent a memoir to the Royal Society, in which he proposed a trigonometric survey connecting the Observatories of Paris and Greenwich for the purpose of better determining the latitude and longitude of the latter. This proposal was accepted and brought to the Anglo-French Survey (1784–1790), the results of which were published in 1791. During the work, he visited England, meeting with other famous astronomers like Pierre Méchain, Adrien-Marie Legendre, and William Herschel at Slough. In 1788, he was also elected a Foreign Honorary Member of the American Academy of Arts and Sciences. He studied the variations of the declination of the Earth's magnetic field in 1792, specifically, the difference between the direction of the magnetic and geographic poles. In this respect, he also invented a compass of declination, which he called *absolute* because it was equipped with a telescope that enables the determination of the geographical meridian by means of astronomical measurements.

As mentioned in the previous section about Cassini III, Jean-Dominique completed the famous "map of France" which was published by the Academy of Sciences in 1793 and served as the basis for the *National Atlas* (1791), depicting the departments of France. During the regime of Roberspierre, his position became intolerable because of his relationship with the monarchy, so he resigned from the directorship of the Observatory on September 6, 1793, but in the following year Cassini was imprisoned. After seven months, he was released and that terrible experience necessitated his retirement to his manor of Thury. Here, he started to help his county by carrying out humanitarian and agricultural projects, writing the "*Mémoires pour servir à l'histoire de l'observatoire de Paris*", published in 1810. The volume contains his *Eloges* (eulogy) of several academicians and the autobiography of his great-grandfather Giovanni Cassini.

He died at the advanced age of 97, on October 18, 1845, in Thury where is buried.

Jean-Dominique married Claude-Marie Louise de La Myre-Mory and had six children: Anne-Cécile, Angélique-Dorothée, Alexandre-Henry-Gabriel, Aglaée-Elisabeth, Aline-Françoise, and Alexis-Jean-Dominique. The latter, and the second of the two sons, died in the battle of Braga in 1809, while the continuation of the "dynasty" is usually attributed to Alexander-Henry-Gabriel, Cassini V. Although this son was the youngest member of the dynasty, he died before his father Jean-Dominique, who then was the last remaining descendant of the Cassini family and left all the scientific books of his ancestors to the Observatory and to the town council of Thury.

Cassini V

Alexandre-Henry-Gabriel

Alexander-Henry-Gabriel was the only son of Jacques-Dominique surviving to adulthood. He was born in the Observatory on May 9, 1781, where he learned the fundamental principles of astronomy, but he preferred to study law, and he had a brilliant career in the magistracy.

He also had a penchant for botany: he published 65 papers and eleven reviews in the *Bulletin des Science* between 1812 and 1821 and named many flowering plants and new genera in the sunflower family (Asteraceae), many of which were from North America. In fact, he did several experiments with new cultures on his family lands in Thury and in 1825 placed the North American taxa of Prenanthes, family Asteraceae, tribe Lactuceae, in the newly-identified genus Nabalus.

Abrotanella, Brachyscome, Carphephorus, Celmisia, Dracopis (coneflower), *Emilia* (tasselflower), *Eurybia, Euthamia* (flat-topped goldenrod), *Facelis, Glebionis, Guizotia* (niger-seed), *Heterotheca* (camphorweed, golden-aster), *Homogyne, Ixeris, Ligularia, Pallenis, Pluchea* (marsh-fleabane), and *Sclerolepis, Youngia.* All these names refer to some genera originally named by Cassini and in 1828 named *Dulgadia hoopesii* for the Scottish naturalist Dugald Stewart (1753–1828). The genus Cassinia was named in his honor by the botanist Robert Brown.

In 1831, Alexander-Henry-Gabriel was appointed a Peer of France, and the Count Alexander-Henry-Gabriel de Cassini died of cholera on April 16, 1832. With him, the Cassini family finally came to an end. He had married Catherine-Elisabeth-Agathe de Riencourt (his cousin), with whom had no children.

Curiosity

As reported earlier, several women were also born in the Cassini dynasty. It is then natural to ask oneself whether, as it often happened in other families, they helped other astronomers in their professions. Famous in this sense are Tycho and Sophie Brahe, Hevelius and his wife (Fig. 20.3), Manfredi and his sisters, Lalande and

Fig. 20.3 Hevelius and his wife Elisabetha making observations, 1673

several friends and relatives, or William Herschel and his sister Caroline, just to mention the most well-known examples in various eras. In the case of the Cassini family, however, there is no trace of their actual involvement. Rather, it is mentioned in Cassini I's memoirs that, during the last two years of his life, he donated a beautiful spinning wheel to Suzanne, his daughter-in-law, to thank her for playing the harpsichord and singing Italian songs.

Other women, coming from other families, were also present in the Observatory at that time: the "*Demoiselles de l'Observatoire*" [the girls of the Observatory] were the daughters of Couplet and of Philippe de La Hire, two other astronomers working in this institution. Both, with their presence, gave solace during the last period of the life of Cassini, the first with some sweet presents like jam, the second by discussing with him his favorite books, because "*elle en est fort capable*" [she is very capable]. The list included the Holy Scriptures, Cicero, Erasmus translated into French, Pliny the Elder translated into Italian, *The Geography* of Ptolemy and of Strabone, La Fontaine's Fables and Ariosto, who he defined: "*un bon talent employé en des choses frivolles*", namely a good talent employed in frivolous things.

Chapter 21
Nephews Maraldi

The Maraldi family that was based in the little village of Candeasco close to Oneglia, in Liguria, became joined with the Cassinis at the time of Cassini I, and over the years it provided several astronomers who collaborated with their famous relatives. These members are also identified with Roman numerals, especially to avoid being confused with each other because of the recurrent cases of homonymies.

Giacomo Filippo Maraldi or Jacques Philippe (Perinaldo 1665—Paris 1729)

Maraldi I

Giacomo Filippo was born in 1665 in Perinaldo, the same village as his uncle Giovanni Domenico Cassini. The exact date is uncertain: for example, on the commemorative plaque in Maraldi Street, it is reported as April 21, but his biographer Fabroni indicates instead September 12, and another biographer, Fontanelle, August 21. Unfortunately, the book of baptisms has been lost, and this controversy cannot be settled. His parents were Francesco Maraldi and Angela Caterina Cassini, the youngest sister of Giovanni Domenico Cassini.

Giacomo Filippo was the first of eight children, and probably he studied at home with a tutor and in some nearby college afterwards. He showed great interest in mathematics and natural philosophy, so his famous uncle called him to Paris. He arrived at the Observatory in 1687, and there he lived and worked the rest of his life, alongside his uncle as his assistant, accompanying him on astronomical observations, in travels, and at scientific meetings.

G. Bernardi, *Giovanni Domenico Cassini*, Springer Biographies,
DOI 10.1007/978-3-319-63468-5_21

He became a member of the Royal Academy of Science in Paris in 1699 and contributed 112 essays about his research and observations, which concerned, for example, comets, variable stars, eclipses, Jupiter, Saturn, and sunspots. He never married and lived in the Observatory with his uncle and then with his cousin and his family (Fig. 21.1). Giacomo Filippo called to Paris his nephew Giovanni Domenico from Perinaldo, as his uncle bearing the same name did with him.

In the Observatory, in addition to his work as assistant astronomer to Cassini I, he devoted himself to personal projects. He made extensive studies of Mars, and his most famous astronomical discovery was that the ice caps on Mars are not exactly located at the geographical poles of the planet. Maraldi I also discovered, in May 1724, that the corona visible during a solar eclipse belongs to the Sun, not to the Moon, he identified R Hydrae as a variable star, and he helped with the survey based on the Paris Meridian.

He is also credited for the first observation, made in 1723, of what is usually referred to as Arago spot or Poisson spot, an observation that was unrecognized until its rediscovery in the early 19th century by Dominique Arago. At the time of Arago's rediscovery of this spot, due to a diffraction effect, it provided convincing evidence of the then-contested wave nature of light.

In his time, astronomers were still using the Bayer catalogue *Uranometria*, published in the early 16th century. The British astronomer Flamsteed was calculating the positions of more than 3000 stars, and Giacomo Filippo worked on his catalog of fixed stars throughout his life. Methodical and constant, he observed on every clear night for over 40 years, a heavy commitment as recalled in his Eulogy by Fontanelle "*...le Géometre les plus laborieux meme presque une vie molle au prix d'un Astronome également occupé da sa science...*" and his health was compromised; in fact, he suffered from frequent stomach pain.

When only a few stars close to the celestial pole were missing to complete his catalog, Giacomo Maraldi mounted a mural circle on the Observatory terrace to observe them better, but his health suddenly worsened, and he died, on December 1, 1729, leaving his endeavor unfinished.

Unfortunately, the catalogue was never published, and only some parts were released and used by the astronomers Delisle and Manfredi for his Ephemeris. Eventually, when, about 70 years later Johan Elert Bode published the *Uranographia*, a large celestial atlas with more than 15,000 stars, the lifetime work of Maraldi became undeniably pointless. In this respect, we have mentioned how the astronomer Delambre was surprised and disappointed that no family member or fellow astronomer had completed the work and had it printed to honor the memory of Maraldi.

Giacomo Filippo Maraldi is also known in mathematics for having obtained the angle of the rhombic dodecahedron shape in 1712, which is still called the Maraldi angle. He is also known by the French version of his name, Jacques Philippe, and as Maraldi I, to distinguish him from the latest of the Maraldi astronomers.

Fig. 21.1 Portrait of Maraldi I on display at the Astronomical Museum of Perinaldo (Author's photo reproduced with the permission of the city of Perinaldo)

Giovanni Domenico Maraldi (1709–1788)

Maraldi II

Giovanni Domenico Maraldi was born in Perinaldo on the April 17, 1709, son of Angela Francesca Allavena and another Giovanni Domenico Maraldi, the younger brother of Giacomo Filippo, or Maraldi I. Giovanni Domenico shared the same name with the famous Cassini I, and he would become an astronomer as well. He studied at home, then in San Remo in the Jesuit College, and finally mathematics, physics, and medicine in Tuscany at Pisa (Fig. 21.2).

The enthusiastic Maraldi II, at the age of 18, in the spring of 1727 came to Paris at his uncle's (Maraldi I) invitation. There he met the French branch of the family: Cassini II, Jacques, and his son. The Observatory was a large building but with only a few rooms, so Giovanni Domenico had to sleep in the wide embrasure of an upper storey window, but for him this was not a big problem, as confirmed by these lines: "*...cette petite cellule était très-conforme à son gout pour la solitude et à son caractère un peu sauvage, qui lui faisait trouver bon de ne pouvoir recevoir q'une seule visit à la fois*" [this little cell was very conformable to his taste for solitude, and to its somewhat savage character, which permitted him to appropriately receive only one visit at a time].

Giovanni Domenico, in the Observatory, was directed by his uncle, Maraldi I, to observe the moons of Jupiter to determine longitudes, observations that would be conducted for over fifty years. He obtained a longitudinal difference between Greenwich and Paris of 9 m 23 s, while the modern value being 9 m 20.93 s.

Unfortunately, his master died two years after his arrival, but his teacher then became Cassini II. It seems that he enjoyed working with him and also with Nicolas Louis de La Caille on a geodetic project: the survey of the coast from Nantes to Bayonne. The results of this work were published under Maraldi's name in 1744. La Caille died in 1762, leaving all his notes to his younger colleague, so the following year he published his catalogue "*Coelum Australe Stelliferum*" of southern stars.

In 1731, he became a member of the Academy of Science, and between 1732 and 1740 he worked with his cousin, Cassini III, on establishing the boundaries of France, as part of the famous project of the *Carte de France*, the Map of France we have talked about in a previous chapter. Maraldi II also carried out observations of Mercury's and Venus' transits and of several comets, calculating their orbits, and, in September 1746 during the observation of one of them, the De Chéseaux comet, he discovered *"une étoile nebuleuse assez clair, qui est composée de plusieurs étoiles"* [a fairly clear nebula star, which is composed of several stars]. It was what we now call a globular cluster, and, eventually, he observed two of them: M15 in Pegasus on September 7 and M2 in Aquarius on the September 11.

He also contributed to the production of the "*Connaissance des temps*" [Knowledge of the Times], a 25-volume publication of ephemeris commissioned by the Academy, which was used by astronomers or sailors.

Maraldi II fell ill at the end of 1763 and he retired in 1772 to his birth place, but here, with his instruments, he continued to work and to send his observations to Paris. He never got married and died on the November 14, 1788, in Perinaldo, where he was buried in the graveyard of the parish church. As a posthumous honor, a lunar crater was named after him in 1935.

Fig. 21.2 Portrait of Maraldi II on display at the Astronomical Museum of Perinaldo (Author's photo reproduced with the permission of the city of Perinaldo)

Giacomo Filippo Maraldi (1746–1797 or 1814)

Maraldi III

Giacomo Filippo was born in Perinaldo on September 13, 1746. He studied in Dolceacqua and in Nice, and then at the University of Turin, where he graduated in medicine like his father. Afterwards, he performed professional internships in several hospitals abroad. Among other pursuits, he was also at the Hotel de Dieu, in Paris, thanks to the support of his uncle Giovanni Domenico Maraldi, remaining there until the spring of 1771, and then accompanied his uncle back to Perinaldo. In this hometown, he practiced medicine, but he had also acquired experience in astronomy, helping Maraldi II with astronomical observations on planetary satellites.

He married the young Angela Caterina Aprosio on April 21, 1776, and they had nine children, but only a few of them survived childhood, which was normal at that time. The first was Maria Angelica Maraldi, born in 1778, who married a man named Giacomo Valotti. The others included: Giacomo Francesco Filippo Maraldi, born in 1779 who died in 1802; Giovanni Domenico Maraldi, born in 1782 and died in 1783; and Angela Francesca Maraldi, who was born in 1783 and married Giovanni Vincenzo Ferraironi. Giovanni Domenico Maraldi, born in 1786, became a medical doctor and lived for 33 years, while Giacomo Bernardo Maraldi, born 1789, survived for only 3 years, and Giacomo Filippo Maraldi, born and died in the same year, 1792. The eighth son, Giacomo Filippo Maraldi, was born in 1793; he became Governor General of Alessandria and died in 1870, and finally Pietro Bernardo Maraldi, was born 1796 and died in 1837.

According to one reconstruction of the Maraldi family by John Lockerbie (http://catnaps.org/cassini/family.html), a fourth Maraldi, mentioned by Jean Baptiste Joseph Delambre in "*Histoire de l'astronomie au dix-huitième siècle*", was sent to Paris in 1797 to study under Joseph Jérôme Lefrançois de Lalande who held the chair of astronomy in the Collège de France for 46 years, but the young Maraldi IV died shortly after his arrival in Paris.

Another interesting note about Maraldi IV appears in *The Monthly Magazine and British Register. Volume 6. Part II for 1798 from July to December inclusive. [R. Phillips. London.] 1798, p. 268.* At the beginning, we can find the "History of Astronomy for the fifth year of the French Republic, read at the opening of the sitting of the College of France, Nov. 15, 1797, by Jerome De Lalande, Director of the Observatory, and Inspector of the College of France." Here we can read that "*Citizen Jacques Philippe Maraldi, the third astronomer of that name, has sent us the observations which he is constantly making at Perinaldo, near Nice. He has done more; he has sent to Paris the eldest of his four sons, aged 18 years, to labour with me in astronomy. I foresee, by his intelligence and assiduity, that Maraldi the fourth will maintain the reputation of his family, and that of the Cassinis their relations, who have been unhappily lost to astronomy since the revolution.*"

John Lockerbie is not able to say with absolute certainty who, among the children of Giacomo Filippo, is the Maraldi IV cited in this report, but the odds are that it was the second one: Giacomo Francesco Filippo who died in 1802. His death, however, is recorded as happening in Perinaldo.

When his uncle Maraldi II died in 1788, Giacomo Filippo, or Maraldi III, inherited all his possession and continued to send to Paris his astronomical observations, which were published in the "*Mémoires de l'Académie*". He died at Menton at the age of 68, and with him ends the dynasty of the Maraldi astronomers, but not the lineage which, as we have seen, had subsequent members who pursued a military career.

Chapter 22
A Bit of Dark, A Bit of Light, A Bit of Space

The general talk of the virtuosi here is about the Comet...
Edmond Halley

By the end of this book, we know many things about the professional and personal life of Cassini and his family. He dealt with an impressive list of astronomical and also non-astronomical problems, collecting many successes and rewards, and leaving an important heritage to future generations. But this is not the whole story yet, and among his activities there are "shadows" among the lights, as well as some curious aspects that are worth mentioning. In this last chapter, therefore, we want to list some of these less known facts: "défaillances", failures, the design of particular objects, and his last project connected to the city of Bologna.

Edmond Halley, just arrived in Paris from England on Christmas Eve 1680, wrote to Hooke "*The general talk of the virtuosi here is about the Comet...*" In fact, at the beginning of November 1680, a comet had appeared, then disappeared and appeared again just before Christmas, showing a very bright tail, causing great concern for people. Every European astronomer observed it, and so did Cassini as well, who judged from the changes in speed and direction of the comet that this one and the previous were two distinct objects, and that the latter was the same observed by Tycho Brahe in 1577. Cassini observed it from the Observatory with Huygens, Picard (returned from his mission to Tycho Island), Richer, Ole Christensen Rømer (who had joined Picard on his return trip), and another prestigious colleague, the astronomer Edmond Halley himself.

At that time, the appearance of a comet, especially of a spectacular one like that of 1680, was a momentous event both for the scientists and the common people. Alan Cook, in "*Edmond Halley, Charting the Heavens and e Seas*" writes: "*Notions of comets at this time were very varied. Was the comet seen at the end of 1680 the same as that seen early in 1681? Newton and Famsteed debated whether the comet passed behind or in front of the Sun between the two apparitions. Of what form were the orbits, straight lines, great circles, parabolas, or ellipses, and did comet reappear after some time, as Cassini suggested*"? Eventually, it was shown that the comets seen in 1680 and 1681 were actually the same object, so Cassini, always too

G. Bernardi, *Giovanni Domenico Cassini*, Springer Biographies,
DOI 10.1007/978-3-319-63468-5_22

impatient and eager to publish, this time collected an "epic fail" and exposed himself to the criticism of the colleagues.

Indeed, as we have already seen, Cassini had not only admirers among academics and astronomers, but also relentless critics. It is possible that his professional success and proximity to the King encouraged this kind of attitude. The most notable example of such critics is the already cited French astronomer Jean-Baptiste Delambre (1749–1822). He wrote and published a monumental work "*Histore de l'Astronomie moderne*" [History of Modern Astronomy] in which he strives to dismantle Cassini's contributions piece by piece.

Another failure of the great astronomer was in the art of making telescopes. We have repeatedly said in this book that the best telescopes at the time were the Campani spyglasses. The French kingdom made several attempts to obtain the secrets of the Italian manufacturer for the exclusive benefit of France, tempting him with fabulous sums, but he always refused to relocate. So Cassini, with the help of his colleague and friend Viviani, ordered four cases of glass material from Tuscany, addressed to Colbert. They took a long time to reach their destination, raising the suspicion that they had fallen in the wrong hands, and, despite the interest of the famous astronomer, eventually all the efforts to make lenses in France failed to yield results comparable to those of Campani.

The French-Italian astronomer had also his share of missed opportunities, and probably the most significant of them is related to the same objects that boosted his career with his transfer to France, namely the satellites of Jupiter. The previously mentioned Danish astronomer Ole Rømer (1644–1710) arrived in Paris with Picard in 1672, when he was still less than 30, and the King made him tutor for the Dauphin, the heir to the throne of France. In the previous years, with Picard on the Heven Island, they had observed 140 eclipses of moons of Jupiter which had to be compared with the same observations made by Cassini in Paris. As we know, the differences between the times of the eclipses gave the differences in longitude between Heven and Paris. Cassini, however, noted some discrepancies in these times, in particular, about the moon Io. The interval between the eclipses became shorter when Earth approached Jupiter and longer when Earth moved away, so he published in August 1676 a work in which he wrote: "*Cette seconde inégalité paraît venir de ce que la lumière emploie quelques temps à venir du satellite jusqu'à nous, et qu'elle met environ dix à onze minutes à parcourir un espace égal au demi-diamètre de l'orbite terrestre*" [this second inequality appears to be due to light taking some time to reach us from the satellite; light seems to take about ten to eleven minutes to cross a distance equal to the half-diameter of the Earth's orbit]. In other words, he attributed a finite value to the speed of light. As reported in an interesting paper by Bobis and Lequeux "*Cassini, Rømer and the velocity of light*", Cassini, in spite of his known eagerness for publication, then abandoned this hypothesis because, in his opinion, there could be alternative explanations for this phenomenon. One of them was based on the hypothesis that the diameter of Jupiter changed periodically in time, which might seem odd in our eyes, but, with the knowledge of the time, could not be refuted. This unusual caution, eventually, made Rømer the first to propose this theory in his famous paper "*Démonstration touchant le mouvement de la lumière trouvé par M. Roemer de l'Académie des sciences*" on December 7 1676.

Cassini was also a very curious and inventive person, as is demonstrated by the various reports he made to the Academy or on the *Journal des sçavans,* the first academic journal published in Europe. Here, on December 27, 1676, we can read the description of a "*Balance arithmétique*" [arithmetic scale] by Cassini, a sort of mechanical ancestor of those we can now find in any supermarket, useful to know the weight and price of goods. Other examples can be found in the astronomical field. In 1667, Cassini proposed a new "*Jovilabe*" or an instrument to represent the motions and configurations of Jupiter's satellites, and in 1681 he presented to the king a new silver planisphere: two overlapping plaques were engraved with the constellations visible from Paris, an ingenious mechanism making it possible to configure the sky at any time of the year. Similarly, he also conceived a heavenly globe for perpetual use, like the modern planetarium, built by the King's engineer Bion.

Cassini, during his last trip to Italy, had been approached by the noble Luigi Ferdinando Marsili, or Marsigli (Fig. 22.1), who had the intention to set up an Institute of Science in Bologna based on the French model. On his return to Paris, the famous astronomer gave his support to this project, providing advice and sending books that would constitute the first Marsilian library. He collaborated on this project also with the young Eustachio Manfredi, professor of astronomy at the University, who also inherited the responsibility of the meridian line in San Petronio and would become the first director of his private and later public observatory in Bologna. Speaking of this city, in 1670 Cassini had requested the citizenship of Bologna, but soon after he left the University to settle permanently in Paris; only in 1702 did Cassini receive the citizenship, after more than 30 years of waiting.

Cassini's last effort was the topographic survey in southern France for the extension of the meridian of Paris discussed in previous chapters. According to the records of the Academy, in the first ten years of the 1700s, this project is connected with his name and also with those of his son and of his nephew Maraldi. This was surely due to his age which necessitated the collaboration of his younger relatives, and, indeed, the travel of 1701 was done with his wife, who took advantage of the thermal baths near Montpellier for her rheumatism. In particular, Maraldi was entrusted with the final studies on the reform of the Gregorian calendar, and their attendance in the Rome meeting. This reform regarded the coincidence of the liturgical year with the astronomical one, so that the spring equinox always fell on March 21 and could fix the Church's moveable celebrations like Easter. As a final celebration of this important achievement, a fine meridian line, similar to that of San Petronio in Bologna, was traced in the Basilica of Santa Maria degli Angeli by the astronomer Bianchini, with the help of Maraldi I (Fig. 22.2).

Fig. 22.1 Portrait of Luigi Ferdinando Marsigli

Cassini-Huygens

Together in the space

Believe it or not, Cassini and Huygens, the two most famous scientists in the Academy of Science of Paris, centuries after their deaths teamed up again for a scientific endeavor in space to explore Saturn and Titan, two objects whose discovery and research is permanently linked with their names, as we previously learned.

With the advent of the space age, it was inevitable that a mission to the "Lord of the Rings", about which Cassini and Hugyens had made important discoveries of particular note with the best instruments of the time, would be organized. Huygens discovered Saturn's rings and, in 1655, its largest moon, Titan. Cassini, instead discovered four more satellites of this planet: Iapetus, Rhea, Tethys, and Dione, and in 1675 he discovered what is known today as the "*Cassini Division*", the largest among the narrow gaps separating the rings. It was then natural that a mission intended to realize the most detailed study of the Saturn system and of Titan in our age was named after these two great scientists.

Fig. 22.2 Diagram of the meridian line in Santa Maria degli Angeli, Rome [From Bianchini, De nummo (1703)]

The Cassini-Huygens mission is the fourth space vehicle that has visited Saturn system, but it is actually made of two probes. The largest one, Cassini, which is also the first one to orbit the planet, and Huygens, a smaller probe built to land on Titan and explore its surface. This mission is the result of a joint effort of the NASA, the European Space Agency ESA, and the Italian Space Agency ASI, and it is probably one of the most complex probes conceived so far (Fig. 22.3).

The decision to build a dedicated probe for the exploration of the largest moon of Saturn is based on the complex environment possessed by this large satellite, which resembles in many ways that of the ancient Earth with its methane-based atmosphere, supposedly similar to the one where life began on our planet.

On the other hand, the system formed by Saturn, its satellites and its rings, is so complex that it required an extremely large and advanced probe, Cassini, able to perform several difficult orbital maneuvers for an extended time.

Launched on 15 October 15 1997, the spacecraft reached its final destination on July 2004. Here it started the exploration of the system and landed Huygens on Titan on January 14, 2005, with Cassini serving as a relay to send the probe's data to Earth. The latter then continued the first phase of the exploration of the system, ending in 2008, exploring the rings and some of its satellites, like Enceladus. After that, the satellite was still "in good shape", which allowed first a two-year mission extension until October 2010 called "Cassini Equinox Mission" and then a second and even longer extension, the "Cassini Solstice Mission" from 2010 to 2017, with

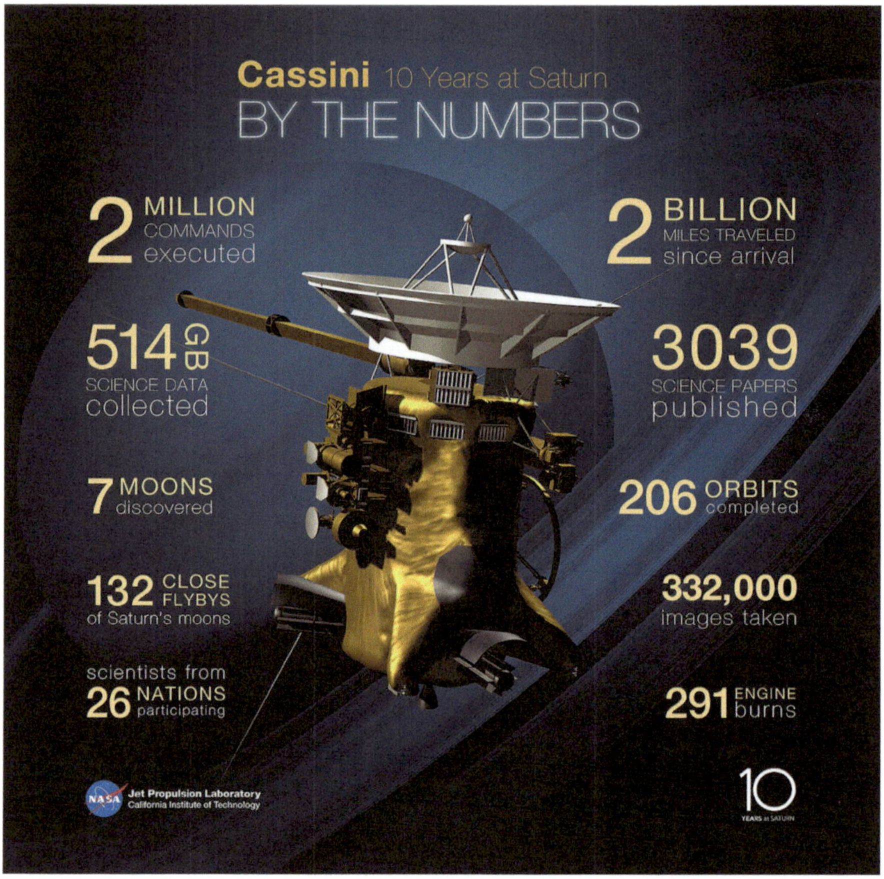

Fig. 22.3 The Cassini probe completed its four-year primary mission in 2008 and went on to perform dozens more flybys of Titan, Enceladus, and Saturn's other icy moons through its 10th anniversary in 2014. The mission continued through 2017 (NASA/JPL-Caltech)

more flybys of Saturn's satellites, including Titan, the observation of storms on the main planet, and other exciting discoveries. Now Cassini has entered into the last phase of its mission, the "Grand Finale", which is specifically dedicated to the rings and the atmosphere of Saturn. We are all waiting for its conclusion on September 15, 2017, with the penetration of the probe into the atmosphere.

Chapter 23
Conclusion

> *...les Astronomes...nous donnes des yeux,*
> *et nous dévoilent la prodigieuse magnificence de ce Monde*
> *Presque uniquement habité par des Aveugles...*
>
> Bernard de Fontanelle

During the Little Ice Age, it was not unusual to see the rivers of the northern Europe freezing. In Paris, the Seine froze in the winter of 1709, and Cassini became seriously ill, but by spring he was almost fully recovered. His wife, Geneviève Delaistre, had died the previous year, on September 17, 1708, aged 55, and this event is reported in Cassini's logbook with the simple words: "... *Madame Cassini est morte aujourdui a 11 h du soir*" [Mrs. Cassini died today at 11 pm] preceded by the values of atmospheric pressure and temperature. But life continues around him with happy events like the birth of a granddaughter, and his family and assistants were all occupied in astronomical work in the Observatory.

He continued to be very active, collaborating with many scientists and, among others, with the young scientist Manfredi and his patron Marsili, whom he had met during his last journey in Bologna, or the Genoese noble Paris Maria Salvago. But even if he had completely recovered from the recent illness, two years later Cassini became completely blind. This did not prevent him from attending the Academy meetings, and he started dictating his memoirs to his secretary, currently preserved at the National Library of Paris, alternating this job with prayer. This important document, which occupied the time of the scientist for his last two years of life, is a manuscript volume of 324 pages, which ends on September 11, 1712, three days before his death.

Already in the fall of 1711, Cassini complained of: "*...une si grande faiblesse que je ne me pouvoit soutenir...*" [...such a great weakness that I could not support myself...] and his faithful nephew assisted him with applications of warm cloths. Cassini was so weak that he had to be supported by two people to go on the terrace of the Observatory; he could not attend the meeting of the Academy anymore (its president, Christiaan Huygens, had died long before in 1695 in the Netherlands); and he began to hear Mass in a private chapel set up inside the Observatory, on a wheelchair.

G. Bernardi, *Giovanni Domenico Cassini*, Springer Biographies,
DOI 10.1007/978-3-319-63468-5_23

Fig. 23.1 Francesco Bianchini holding the eyepiece mount of an aerial telescope

In his memoirs on June 6, 1712, it is reported that Cassini began to be read the history of the Roman politician Appius Claudius Ceacus [the blind], also known for his political speech in Latin "*quisque faber suae fortunae*" [each man is the architect of his own fortune], then Homer "*qui fit beaux ouvrages dans son Aveuglement*" [who made fine works in his blindness] and concluding with a list of "*Les astronomes célèbres du Siècle passé qui ont perdu la vue dans leurs dernière années dont Galilée, le Comte ... Viviani et Montanari*" [the famous astronomers of the past century who lost sight in their last years like Galileo, Count… Viviani, and Montanari].

In late August of the same year, the astronomer Francesco Bianchini arrived from Rome, for the last professional visit to the old scientist (Fig. 23.1). Bianchini brought a model of the celestial sphere of the Farnese Atlas which he wanted to compare with the "*Globe perpetual*" [perpetual globe] built years before by Cassini. Unfortunately, because of his condition, Cassini could not accompany his guest to visit the Observatory, but as always he showed a sincere interest in his research.

The end is approaching: on Sunday, September 11, 1712, Cassini did not come out of his room, but he enjoyed the visit of friends and relatives, as he reported: "*Mademoiselle De la Hire vint la matin prendre avec moy et avec Mr. Maraldi une tasse de chocolat…Le soir, je fis continuer la lecture de la Bible.*" [Miss de la Hire came in the morning to take with me and Mr. Maraldi a cup of chocolate… In the evening I continued the reading of the Bible]. These are the last words transcribed by Cassini. Three days later arrives: "*…la mort, 14 septembre; on peut dire qu'il s'endormit dans le Seigneur.*" […the death, September 14; it can be said that he fell asleep by the Lord]. This sentence was added by Cassini IV at the end of the manuscript, but nobody is reporting the event in his Journal of Observations. On that day, we can only read that there blew a south wind.

The news is published in the Gazette of Bern and from there it spreads across Europe and is followed by many commemorations, celebratory orations, and a touching endnote of a letter from his lifelong friend Salvago to Manfredi: "*… [Cassini] non morirà mai il suo nome, sino a che vi siano cieli e stelle da contemplare…*" […his name will never die as long as there are heavens and stars to behold].

On September 16, 1712, Cassini was buried in the Church of Saint Jacques du Haut-Pas, and on the tombstone is written simply: *Jean Dominique CASSINI astronome mort le 14 September 1712*, [Jean Dominique Cassini astronomer died on September 14, 1712]. Now his remains are no longer there, because in 1850 they were transferred to the catacombs of Paris.

Exactly two months after his passing, Bernard de Fontanelle remembered Cassini in the Academy with these words: "*…les Astronomes…nous donnes des yeux, et nous dévoilent la prodigieuse magnificence de ce Monde Presque uniquement habité par des Aveugles…*" ["… the Astronomers… give us eyes, and reveal to us the prodigious magnificence of this world, inhabited almost uniquely by blind people…"].

The following is an incomplete list of his published works that we have seen in detail in this book, but this is only a summary of the contributions that this scientist made to enlighten humanity (Fig. 23.2). A more detailed list, prepared by Cassini IV, is reported in Appendix E.

- *Ad serenissimum principem Franciscum Estensem Mutinae Ducem. Io. Dominicus Cassinus Genevensis In Bononiensi archigymnasio publicus astronomiae professor. De cometa anni 1652 et 1653*, Mutinae 1653.
- *Ephemeris prima motus cometae novissimi mense aprili 1665* […] *ad collationem cum observationibus ex primis ejus apparentiis deducta ad horam 4. post mediam noctem Romae, a Joanne Dominico Cassino*, Romae 1665.

- *Lettere astronomiche di Gio. Domenico Cassini al signor abbate Ottavio Falconieri sopra il confronto di alcune osservazioni delle comete di quest'anno 1665*, Roma 1665.
- *De periodo quotidianae revolutionis Martis*, Bononiae 1666.
- *Martis circa axem proprium revolubilis observationes Bononiae a Jo. Dominico Cassino habitae*, Bononiae 1666.
- *Ephemerides bononienses mediceorum syderum ex hypothesibus, et tabulis Io: Dominici Cassini almi Bononiensis archigymnasij astronomi ad observationum opportunitates praemostrandas deductae. Ad eminentissimum principem Iacobum S.R.E. Cardinalem Rospigliosum*, Bononiae 1668.
- *Spina celeste meteora, osservata in Bologna, in mese di marzo 1668, da Gio. Domenico Cassini*, Bologna 1668.
- *Découverte de deux nouvelles planètes autour de Saturne*, Paris 1673.
- *Abregé des observations et des reflexions sur la comete qui a paru au mois de decembre 1680, et aux mois de janvier, fevrier et mars de cette année 1681. Presenté au Roy par Mr. Cassini*, Paris 1681.
- *Le Neptune françois ou Atlas nouveau des cartes marines levées et gravées par l'ordre exprès du Roy pour l'usage de ses armées de mer, dans lequel on voit la description exacte de toutes les côtes de la mer Océane et de la mer Baltique, depuis la Norwège jusques au détroit de Gibraltar. Reveu et mis en ordre par les sieurs Charles Pène, Cassini, et autres*, Paris 1693.

Fig. 23.2 Detail of the statue of Jean-Dominique Cassini at the Observatory of Paris showing the main achievements of the astronomer (Author's photo)

- *Recueil d'observations faites en plusieurs voyages par ordre de sa Majesté pour perfectionner l'astronomie et la geographie. Avec divers traitez astronomiques. Par Messieurs de l'Academie Royale des Sciences*, Paris 1693.
- *Mémoires pour servir a l'histoire des sciences et a celle de l'Observatoire Royal de Paris, suivis de la vie de J.-D. Cassini, écrite par lui-même, et des éloges de plusieurs académiciens morts pendant la Révolution*, Paris 1810.

Certainly many of these researches and works would not have happened if Cassini had not benefitted from the right conditions: he could enjoy the full support of the Academy of Science of Paris, Colbert, and Louis XIV, eventually making it possible him to explore in the directions he preferred, perhaps the greatest aspiration of a modern scientist.

Appendix A

This astronomical poem "*Frammenti di Cosmografia*" (Fragments of Cosmography) by Giovanni Cassini is written in Italian. Apparently, it is dedicated to the Christina, the Queen of Sweden, but it is also very peculiar because at the time of its writing the theory of celestial motions governed by Newton's law of universal gravitation had not been published yet.

The "*Imitation libre du fragment precedent*", or a free imitation of a previous piece, in French language, follows.

Frammenti di Cosmografia

PRENDO a ridurre a mente in brevi detti
Del mondo la struttura, e gli elementi,
I siti delle stelle, e i movimenti,
I congressi, gli eclissi, e i varj aspetti.

STUDIO degno di Voi spirto gentile,
Ch' il doppio vol degli occhi al cielo ergete,
E nata gli astri a contemplar, chiudete
Nel più bel seno un' animo virile.

FORSÈ nel sen chiudea spirito tale
La degna fondatrice di Cartago,
Che dell' ospite Enea l'orecchio pago
Fè di tal canto al talamo regale.

G. Bernardi, *Giovanni Domenico Cassini*, Springer Biographies,
DOI 10.1007/978-3-319-63468-5

UDIR fè d'Iopa in sù l'aurata cetra
Ciò, che insegnar soleva il grande Atlante,
I movimenti della luna errante,
E negli eclissi spuntò il sol nell'etra:

ONDE son le recondite cagioni
Dell' umana progenie, e delle sfere,
Onde cadon le pioggie, e il fulmin fere,
Arturo, l'Iadi, e i Gemini Trioni:

PER qual cagione il sole in sì poche ore
A l'Oceano arrivi i dì del verno,
E perchè poi nell'emisfero inferno
Faci in la notte allor tante dimore.

QUESTI pur del vostr' animo regale
Sono i più grati, e placidi diporti,
A quant' altri pon dar splendide corti,
Quest' unico diletto in Voi prevale.

CIO che in ciel già leggeste, e sù le carte
Quì descritto vedete in rozzi versi,
D'uopo non è che sian limati e tersi,
Che la materia qui supplisce all'arte.

RICUSA ogni ornamento alta dottrina,
Che per se sola assai diletta, e piace.
D'ogn' altra musa Uranìa più verace
Fugge ogni vana pompa, e peregrina.

TULLIO, che quanto a declamar fecondo
Tanto ebbe in poesìa lo stile ingrato,
Tradurre in versi osò l'opre d'Arato,
Che tutto descriveano il cielo, e il mondo.

AVRO' splendor bastante, Uranìa mia,
Dal nome vostro, a cui del cielo il canto
Con più giusta ragione offrir mi vanto,
Ch' Alessandro la sfera a Laudomia.

MONDO diciam quest' ordine di cose,
Ch' alla vista mortal si rappresenta,
Di cui l'orbe terren centro diventa,
Ove Dio l'uomo ad osservar ripose.

Così se d'abitar ci fusse dato
O luna, o corpo tale in se sospeso,
Quel per centre del mondo avriano preso
L'ordine tutto intorno a lui formato.

A noi nasce quest' ordine del mondo
Per conseguenza d'ottica ragione,
Che quanto lungi discopriam, dispone
Intorno al'l' occhio nostro in giro tondo.

PERCIO' pigliam dell'occhio nostro il centro
E alla maggior che concepiam distanza,
La sfcra descriviam per ampia stanza,
Che l'univcrso tutto includa dentro.

SFERA, che ampia formiam, quanto a noi pare
Pur che capace sia di tutto il mondo,
In guisa tale egli riman rotondo,
E nel mezzo contien la terra e il mare.

QUINDI è, che gli astri tutti, o sian rimoti
Dall' occhio e dalla terra, o sian vicini
Là negli estremi sferici confini
L'occhio trasporti, e l'arte industre noti.

Ivi quella distanza, o positura
Diciam fra loro stesse aver le stelle,
Ch' anno i lochi lassù, che queste e quelle
Occupan nella sferica figura.

Ivi le stelle allor diciam congiunte,
Saturno per esempio a Giove, a Marte,
Quando ci copron la medema parte,
Benchè sempre fra lor siano disgiunte.

MA perchè troppo par vario e vagante
Preso dall' occhio il centro della sfera,
Prendiamo il centro della terra intiera,
Ondè ogni occhio abbiam quasi equidistante.

PERCIÔ diciamo positura vera,
Vero diciamo d'ogni stella il sito,
Che l'occhio dal centro della terra unito
Vedrebbe aver nella suprema sfera.

Di terra, d'acqua, e d'aria insieme poste
Si compone la sfera elementare,
In se sospesa, e atta a conservare
Le cose corruttibili, e composte.

NELL' aria distinguiam tre regioni,
Nell' infima respiran gli animali,
Nella media le nubi anno i natali,
E le pioggie, e le nevi, e i lampi, e i tuoni.

NELLA suprema quanti veggiam fuochi
Ogni notte volar serena e oscura,
Se celeste non è la lor natura,
Ne ammette il ciel si momentanei giochi.

Noi sin lassù sappiam che l'aria ascende
Dove salir può mai fumo o vapore,
Alto a prender del sol dubio splendore,
Onde or la prima sera, or l'alba splende.

DEGLI alti monti poco più sublime
E di quest' aria la maggior' altezza,
Mentre in breve a lei manca ogni chiarezza
Quando de' monti il sol lascia le cime.

GIA sopra l'aria, e sotto il ciel le scuole
Una sfera introdussero di foco,
Per dar a un elemento il quarto loco,
Che per aria aspirar in alto suole.

E, diceano, la terra in sommo greve,
E perciò va del mondo alle parti ime,
Segue l'acqua, e poi l'aria, e più sublime
Di tutti è 'l foco, perchè in sommo è lieve.

MA non an tutti gli elementi il loco
Distinto in regioni ime e superne,
Che della terra ancor nelle caverne
Ora acqua, ora aria, ora troviamo foco.

GIA l'acqua non sormonta il continente
Ambiziosa di compir la sfera,
Quando in diluvio universal non pera,
Come sotto a Noè l'umana gente.

PERCHÉ dunque coprir dee l'aria tutta
Una sfera vastissima di foco,
Se non quando la terra in ogni loco
Da incendio universal sarà distrutta.

CHI staccato giammai da face o pira
Vidde foco spiccar per l'aria un salto,
E lucido inestinto andar in alto
Sin alla sfera ove diciam ch' aspira?

CONGEDIAM, che sian quattro gli elementi
De' corpi misti, terra, acqua, aria, e foco,
Ma questo fugge, e non ha proprio loco,
E sono gli altri tre sol permanenti.

TANTO s' estingue qui, quanto s'accende
Di foco nella sfera elementare,
E in van sù l'aria osiamo collocare
Foco inestinto, che non arde o splende.

SOLO veggiamo nell'oscure notti
Volar fochi veloci, e repentini,
Ch' esser dentro gli aerj confiai
Con probabil ragion stiiuano i dotti.

ALLA scorta fedel del senso retto
La prudente ragion si dona e cede,
Nè opra contraria a quel, ch' ogn' occhio vede,
In natura introdur dee l'intelletto.

Lo spazio oltre la sfera elementare,
Quando voto non sia, d'etere è pieno
Limpido, permeabile, e sereno,
Clic riflessi non può quaggiù mandare.

QUESTA di tutto il cielo è la natura,
Che le stelle sospese in se contiene
Con pace di Stagira, che sostiene
La materia del ciel solida e dura.

OGNI stella può star libera, e sciolta
In mezzo al liquido etereo sospesa,
Come la terra ch' al suo centro pesa,
Si libra in aria tutta in se raccolta.

DA questo illustre esempio il modo è certo
Di star nel mondo i corpi in se raccolti,
Noi da questo uno argomentiam di molti
Ogn' altro, che fmgiamo, è vano, e incerto.

MENTRE da spazioso etereo campo
All' aria nostra viene a far passagio,
Piegasi alquanto d'ogni stella il raggio,
E refratto divien per tale inciampo.

Cosi mentre dall' aria all' acqua passa
II raggio visual si suol piegare,
Onde che rotto in acqua il remo appare,
Ed ogni cosa immersa appar men bassa.

Di tutti li astri un accidente tale
Varia alquanto l'aspetto e positura,
Indi del sol la sferica figura
Al nascer, al cader si mostra ovale.

MOSTRA l'esperienza, e la ragione
Che tale effetto in molta altezza è poco,
Ma quanto basso più del cielo è il loco,
Ivi tanto è maggior refrazzione.

AL nascer, al cader si mostra il sole
Tanto alto più per questo solo effetto,
Quanto del sole istesso il chiaro aspetto
Nello spazio celeste occupar suole.

MA nel salir scema la frode, e tosto
A l'insensibil quasi si riduce,
E infin nel vero luogo il sol riluce
Giunto del ciel al più sublime posto.

ALTROVE non troviam che c' interrompa
Del sole, o d'altra stella il puro raggio,
Segno assai certo, che non fa passaggio
Per foco, o cielo sodo, in cui si rompa.

Imitation Libre

Du Fragment Precedent

J'entreprends de réduire à de courtes leçons
La science qui règle et fixe les saisons,
Qui du vaste univers enseignant la structure,
Et des astres errans la marche toujours sûre,
Aux regards des humains atteste la grandeur
Des merveilles du monde et de son créateur.

Daignez être en ce jour la muse qui m'inspire
Sur un si grand sujet tout ce que je dois dire,
Princesse, dont l'esprit, le génie et les yeux
Semblent être formés pour contempler les.cieux,
Et qui réunissez ce qui manque à toute autre,
La force de mon sexe et les grâces du vôtre.

Telle autrefois Didon: on la vit comme vous,
Jeune et reine, annoncer les plus sublimes goûts,
Quand, pour former les nœuds d'une amoureuse chaîne,
Après un doux accueil la tendre souveraine
Dans de nobles chansons crut trouver les moyens
D'enflammer à son gré le héros des Troyens.

Iopas, devant lui, répéta sur sa lyre
Les secrets dont Atlas avait daigne l'instruire:
Du soleil, de la lune il décrivit le cours,
Expliqua, dans ses chants, comment dans leur concours,
De l'ignorant vulgaire étonnant la pensée,
La lumière de l'un par l'autre est effacée;
Pourquoi, pendant l'hiver, dans le sein de Thétis,
Phébus plonge sitôt ses rayons amortis.

Et pourquoi, dans l'été, la nuit est toujours lente
A tempérer de l'air la chaleur accablante.
Puis, discourant sur l'homme et sur les animaux,
Sur la pluie et la foudre et les autres fléaux,
Aux Troyens étonnés expliqua toutes choses,
Et parla savamment des effets et des causes.

VOTRE esprit, insensible à tout autre plaisir,
De ces mêmes objets aime à s'entretenir,
Princesse; ainsi mes vers oseront vous redire
Ce que déjà vos yeux dans le ciel ont su lire.
Vous prêterez l'oreille à mes faibles accens,
En faveur de sujets nobles, intéressans.

TRAITONS sans ornement une belle matière;
La modeste Uranie a toujours droit de plaire.
L'aigle des orateurs, l'éloquent Tullius,
En vers peu cadencés traduisit Aratus.
Pour moi, de votre nom j'ornerai mon ouvrage;
Du plus heureux succès il m'offre le présage.

PARCOURONS tous les corps et les objets divers
Que notre œil aperçoit dans ce vaste univers.
Sur la terre, placé par le souverain être,
L'homme, du monde entier d'abord se croit le maître,
Et pense, dans l'erreur dont l'orgueil est l'appui,
Que tout ce qui se meut se meut amour de lui.
Dans cette illusion, ses yeux, il faut le dire,
Avec sa vanité concourent à l'induire.
Trompé par son organe, il rapporte toujours
Des astres éloignés et la place, et le cours,
Au fond plus reculé d'une lointaine sphère
Dont le centre est au point d'où l'œil les considère.
L'optique ainsi le veut: il est par conséquent
Pour chaque observateur un centre différent;
Il fallait cependant choisir uu ternie unique,

Centre fixe et consent du système physique;
C'est celui de la terre, il fut donc arrêté
Que supposant un œil en ce point transporté,
Les mouvemens des corps, leur véritable place,
Seraient ceux vus du centre et non de la surface,
D'où s'offrent à nos yeux mille aspects différens,
Que pour mieux distinguer nous nommons apparent.

REVENONS sur le globe où l'eau, l'air et la terre,
Composent, nous dit-on, la Sphère élémentaire,
La sphère qui nourrit et renferme en son sein
Animaux, végétaux et tout le genre humain.
L'air tient enveloppé dans sa région basse
Tout ce qui de ce globe habite la surface.
Les nuages, plus haut, la pluie et les éclairs,
Occupent ce milieu qu'on appelle les Airs.

Dans une région plus élevée encore
Naissent, brillent soudain les feux, le météore
Qui, dans la nuit obscure, aux voyageurs surpris,
Semblent être les jeux de célestes esprits.
Mais ces trois régions qui forment l'atmosphère
Ne sauraient excéder des monts la tête altière,
Car dès qu'à son couchant, Phébus de ses rayons,
A cessé d'éclairer la cime de ces monts,
Bientôt l'air obscurci perd aussi sa lumière,
Et la nuit se répand sur tout cet hémisphère.

QUE dirons-nous ici de cette opinion
Qui d'une quatrième et haute région
Augmente, ou, pour mieux dire, embrase l'atmosphère,
En y plaçant du feu la subtile matière
Qui, l'un sur l'autre assis, veut que chaque élément
Dans un ordre constant reste séparément.
Il n'en est pas ainsi: loin de cet ordre étrange,
Tout atteste ici bas un utile mélange;
Dans le sein de la terre, avec tous les métaux,
Se trouvent combinés le feu, l'air et les eaux.
Vit-on, vit-on jamais la brillante étincelle,
Sortant en pétillant du feu qui la recele,
Vers le plus haut des airs s'élancer comme un trait,
Où la matière ignée alors l'attirerait?
Non: gardons-nous ainsi d'assigner une place
A l'élément fougueux qui franchit tout espace,
Qui répandu partout, partout vivifiant,
Circule en tous les corps en s'y modifiant:
Là, se développant, il dévore, il consume;
Ici près il s'éteint, plus loin il se rallume;

Caché dans les cailloux, il brille dans les airs,
Et son activité remplit tout l'univers.

DES astres jusqu'à nous si l'on n'admet le vide,
Au moins n'existe-t-il qiùm très-subtil fluide,
Un air raréfié, si clair, si transparent,
Qu'il n'offre aucun obstacle aux corps en mouvement,
Et laisse un libre cours aux rayons de lumière.
Aristote, il est vrai, d'opinion contraire,
Voulait qu'un corps solide emplît le -firmament;
Mais qui pourrait admettre un pareil sentiment?

CHAQUE étoile au milieu de l'immense étendue,
Dans le fluide éther, librement suspendue,
Se soutient: c'est ainsi que la terre et ses eaur,
Ses pierres, ses forêts, l'homme, les animaux,
Habitans de son sein comme de sa surface,
Vers un centre commun tendent tous par leur masse,
Se pressent l'un sur l'autre, et d'invisibles nœuds
Les tenant réunis, ils composent entr'eux
Sous la forme arrondie un grand tout de matière
Qui nage enveloppé du liquide atmosphère.
De la même façon concevez tous les corps

Suspendus dans les cieux: et par de vains efforts,
Gardez-vous d'enfanter quelque nouveau système,
Pour nous inconcevable et peu clair pour vous-même.

DES astres lumineux les rayons à nos yeux
Doivent pour arriver traverser deux milieux;
L'un est l'éther, qui n'offre aucune résistance;
L'autre est notre atmosphère, et celui-ci plus dense
Par le rayon heurté, dans ce choc singulier,

Le détourne avec force et l'oblige à plier.
Ainsi j'ai vu cent fois, dessus l'humide plage,
La chaloupe au moment de quitter le rivage;
Les bras tendus, l'œil fixe, à l'aspect du signal
Les rameurs se courbant d'un mouvement égal;
Chaque rame à la fois se soulève, retombe,
Et paraît se briser en se plongeant dans l'onde.
De la réfraction tel est l'effet trompeur;
Il change des objets la forme et la hauteur.
La lune à l'horizon devient plate, inégale,
Et du soleil couchant la figure est ovale;
Tous les astres enfin, bien loin d'être aperçus
Brillans dans leurs vrais lieux, paraissent au-dessus;
Moins il sont élevés, plus fausse est l'apparence;
A de grandes hauteurs, nulle est la différence;
De là, par un effet heureusement produit,
Le jour devient plus long aux dépens de la nuit.

Ainsi l'illusion, fatale à la science;
Doit toujours sur nos sens nous mettre en defiance:
La vérité sans cesse, à nos yeux soustrait,
Si nous ne démêlons l'apparence du vrai.

Appendix B

The following lines in French are the last reported in the manuscript of the anecdotes of the life of Giovanni Domenico Cassini and are all dedicated to the construction of the Paris Observatory. It is interesting to note the detailed analysis of the structure for astronomical use:

> Le bâtiment de l'Observatoire, que le Roi faisait construire pour les observations astronomiques, était élevé au premier étage lorsque j'arrivai. Les quatre murailles principales avaient été dressées exactement aux quatre principales régions du monde. Mais les trois tours avancées que l'on ajoutait à l'angle oriental et occidental du eôté du midi et au milieu de la face septentrionale, me parurent empêcher l'usage important qu'on aurait pu faire de ces murailles, en y appliquant quatre grands quarts de cercle capables par leur grandeur de marquer distinctement, non-seulement les minutes, mai» même les secondes; car j'aurais voulu que le bâtiment même de l'Observatoire eût été un grand instrument: ce que l'on ne peut pas faire à cause de ces tours qui, d'ailleurs, étant octogones, n'ont que de petits flancs coupés de portes et de fenêtres.
>
> C'est pourquoi je proposai d'abord qu'on n'élevât ces tours que jusqu'au second étage, et qu'au-dessus on bâtit une grande salle carrée, avec un corridor découvert tout à l'entour, pour l'usage dont je viens de parler. Je trouvais aussi que c'était une grande incommodité de n'avoir pas dans l'Observatoire une seule grande salle d'où l'on pût voir le ciel de tous cotés, de sorte qu'on n'y pouvait pas suivre d'un même lieu le cours entierdu soleil et des autres astres, d'orient en occident, ni les observer avec le même instrument sans le transporter d'une tour à l'autre. Une grande salie me paraissait encore nécessaire pour avoir la commodité d'y faire entrer le soleil par un trou et pouvoir faire sur le plancher la description du chemin journalier de l'image du soleil; ce qui devait servir, non-seulement d'un cadran vaste et exact, mais aussi pour observer les variations que les réfractions peuvent causer aux différentes heures du jour, et celles qui ont lieu dans le mouvement annuel.
>
> Mais ceux qui avaient travaillé au dessin de l'Observatoire opinaient de l'exécuter conformément au premier plan qui en avait été proposé; et ce fut en vain que je fis mes représentations à cet égard et bien d'autres encore. M. de Colbert vint même inutilement à l'Observatoire pour appuyer mon projet. On suivit donc les premiers plans; les tours et la grande salle furent élevées à la même hauteur; au milieu de la façade méridionale on laissa une petite fenêtre ou ouverture qui donnait au haut de la grande salle, et l'on projeta de tirer sur le pavé, non-seulement la ligne méridienne, mais encore les lignes horaires. Comme l'on craignait que le bâtiment nouveau ne fût sujet à quelque changement, ce qui avait déjà eu lieu dans la partie orientale, et qui avait obligé à reprendre les fondemens plus bas; on différa de paver la grande salle jusqu'à ce que tout effet pût être passé.

G. Bernardi, *Giovanni Domenico Cassini*, Springer Biographies,
DOI 10.1007/978-3-319-63468-5

On proposa de couvrir la grande salle d'une plate-forme bien solide, sur laquelle on pourrait élever un pavillon carré isolé pour servir à l'usage que j'avais proposé, c'est-à-dire, pour pouvoir apercevoir du même lieu tout le ciel et suivre avec le même instrument et de la même place le cours entier d'un astre. Il fut aussi arrêté que la tour septentrionale ne serait pas octogone, comme on l'avait d'abord projeté, mais qu'elle- serait carrée, pour avoir une plus grande face au septentrion. Je' proposai aussi que cette tour septentrionale fût terminée en haut par une salle ouverte par deux fenêtres, l'une orientale et l'autre occidentale, et par une porte méridionale, et que le toit fût percé d'une ouverture ronde, recouverte d'une plaque de cuivre qu'on pût ouvrir et fermer pour l'usage des observations au zénith à l'abri du vent. Cette salle fut depuis appelée le petit Observatoire.

La tour orientale fut laissée entièrement découverte pour le même usage, et on y laissa dans la façade septentrionale une longue fente qui a servi à recevoir et à élever à diverses hauteurs de grands verres objectifs avec lesquels on a découvert le plus petit satellite de Saturne. La grande salle méridienne fut couverte d'une voûte un peu plus élevée que celle de la tour occidentale', au-dessus de celle-ci on laissa un espace creux propre à recevoir un grand hémisphère con cave pour pouvoir y observer le cours journalier du soleil par le moyen de l'ombre d'une boule élevée à son centre; c'est l'instrument appelé par les anciens scaphe. On y devait marquer par observation immédiate les traces journalières de l'image du soleil dans les solstices, comparées à celles des autres jours de l'année, affectées des différentes réfractions.

Ces traces auraient été divisées par des points horaires à l'aide d'une pendule, et auraient fait connaître l'inégalité des arcs horaires causée par les réfractions des rayons solaires. En attendant la construction d'un semblable instrument, je fis placer dans ce lieu enfoncé un grand quart de cercle construit par Gosselin, et divisé avec soin par Lebas. Un coup de vent terrible le renversa et le rendit inutile aux observations. On plaça depuis dans le même lieu des vases d'étain pour observer la quantité de la pluie en divers tems de l'année et son évaporation; M. Sédileau, après avoir suivi pendant quelques années et publié ces observations, enleva ces vases pour d'autres usages. Toutes les voûtes de l'Observatoire furent percées dans le même axe par un trou rond qui répond à un puits contenant un escalier spiral qui descend au fond des caves de l'Observatoire, dont les fondemens sont aussi profonds que son élévation sur le terrain.

Ce puits sert de grand instrument pour l'observation des étoiles fixes proche le zénith; il sert aussi pour mesurer le tems de la chute des corps qu'on laisse tomber des divers étages de l'Observatoire. L'appui de ce degré spiral a servi aussi à soutenir de grands thermomètres d'eau, dont on a observé les variations en divers tems. Les caves de l'Observatoire font aussi voir que le thermomètre n'y souffre pas de variation sensible depuis la plus grande chaleur de l'été jusqu'au plus grand froid de l'hiver; de sorte que l'air de ces caves peut passer pour tempéré et servir à régler les thermomètres. Un peu à l'est au-devant de la porte de la façade méridionale de l'Observatoire, laquelle est élévée d'un étage plus haut que celle de la façade septentrionale, il y a un autre puits couvert d'une pierre, au milieu de laquelle j'ai fait pratiquer une ouverture qui répond aux caves, et que l'on peut ouvrir et fermer pour faire les mêmes expériences et les mêmes observations qu'à l'escalier des caves où l'on est trop exposé à être troublé par les curieux.

La porte méridionale donne sur une grande terrasse où l'on plante des mâts qui servent à élever de longues lunettes. On y a depuis transporté une tour de bois qui était autrefois à Marli, où elle servait à élever les eaux de la Seine qui vont à Versailles; elle sert présentement à élever des verres objectifs à des hauteurs beaucoup plus grandes que celle du bâtiment de l'Observatoire. Cette terrasse est soutenue du côté d'occident par une forte muraille dressée sur la ligne méridienne; une pareille muraille doit s'élever à l'orient et au midi; mais cet ouvrage n'a pas été achevé, parce que ce qui existe suffit pour les observations…

Ici se termine le Manuscrit des Anecdotes de la Vie de J.D. Cassini, rapportées par lui-même.

(The manuscript of the Anecdotes of the Life of J.D. Cassini ends here, as reported by himself).

Appendix C

In the following lines, in French, Cassini presents his theory of sunspots:

> Selon ma théorie, qui représente assez exactement le mouvement des taches, j'établis que ces taches sont sur la surface sphérique du soleil, qu'elles décrivent des cercles parallèles autour des deux pôles, élevés sur l'orbite du movement annuel de 7 degrés et demi, et qu'elles font leurs revolutions dans un tems à peu près égal à celui de la révolution périodique de la lune autour de la terre. Mais elles ne sont pastoujours visibles par nos lunettes dans leurs retours; ce quipeut s'expliquer en diverses manières.
>
> J'imagine que comme le globe de la terre est composé de deux matières, l'une solide comme le continent, l'autre fluide comme les mers, de même le soleil pourrait être composé de deux matières analogues à celles du globe terrestre, dont la solide serait opaque, et la liquide serait la matière de la lumière qui couvre la plus grande partie de la matière opaque, laissant seulement en quelques endroits des pointes comme sont celles de quelques rochers, et qui constituent les taches apparentes. Il y a sans doute, comme dans nos mers, des flux et reflux qui élèvent tantôt plus, tantôt moins cette matière lumineuse; ce qui fait augmenter ou diminuer l'apparence des taches et les transforme en diverses figures en peu de tems.
>
> Celles que nous observâmes au commencement formaient d'abord la figure d'un scorpion avec ses pattes et sa queue. Un peu après, celte partie s'est détachée et a formé des taches plus petites séparées les unes des autres. Elles étaient enveloppées d'une espèce de nébulosité, qui repré sentait à notre imagination les tourbillons qui se forment autour des pointes de rochers par les marées. Il se pourrait faire aussi que comme dans le globe de la terre il y a des volcans qui en certains tems jettent des flammes et des cendres autour d'eux, de même il y en eût dans le soleil.
>
> Ce que nous avons observé particulièrement, c'est que plusieurs taches du soleil, dont nous avions déterminé la situation à l'égard de ses pôles, sont revenues quelque tems après dans la même partie de la surface du soleil, à peu près comme le Vésuve, vu du même endroit au ciel et venant à s'enflammer, paraîtrait de nouveau dans le disque de la terre au même point où il aurait paru auparavant à l'égard des pôles de la terre, avec la même latitude et longitude géographique déterminée dans les révolutions faites après la première apparition: ce qui rend mes conjectures aussi vraisemblables que celles du retour des mêmes planètes au même lieu du ciel après un nombre de révolutions; car ce n'est que par ce moyen que les

G. Bernardi, *Giovanni Domenico Cassini*, Springer Biographies,
DOI 10.1007/978-3-319-63468-5

anciens ont trouvé, par exemple, que Mercure, après avoir cessé de paraître pendant plusieurs révolutions, a été trouvé à son retour pour le même astre; et que Phosphorus et Hesperus, qui anciennement étaient censées être deux étoiles différentes, ont été reconnues pour la même planète Vénus.

Quelques observateurs ont pris les taches du soleil pour des planètes. Tarde leur a donné le nom de Sydera Borbonia. On peut juger, par ce que nous venons de dire, du peu de fondemens de cette hypothèse.

Appendix D

L'éloge de M. Maraldi, suivant l'usage ordinaire, eut dû être fait par le secrétaire perpétuel de l'Académie royale des Sciences, et prononcé dans une des séances publiques de l'année 1789. Mais la révolution survint à cette époque. Les affaires politiques entraînant M. de Condorcet hors de sa sphère, enlevèrent l'Académie un secrétaire éloquent, un géomètre profond, un de ses membres les plus distingués. Il n'y eut point d'éloges pour M. Maraldi. H en méritait un ret j'ai cru devoir réparer cette omission. Parent et élève de cet estimable académicien, ayant vécu long-tems avec lui, j'ai eu l'avantage de le connaître très-particulièrement, et j'ai le droit d'invoquer l'amitié et la reconnaissance pour me prêter leur langage et m'apprendre à le louer dignement. J'ai cru devoir rendre le même hommage à la mémoire de M. Le Gentil et de M. le président de Saron, dont les éloges n'avaient point encore été faits, et se trouveront à la suite de celui-ci.

ÉLOGE DE M. MARALDI

Jean-Dominique Maraldi, né à Perinaldo le 17 avril 1709, était neveu de Jacques-Philippe Maraldi, membre de l'Académie royale des Sciences, que Jean-Dominique Cassini, son oncle, avait fait venir auprès de lui pour l'instruire dans l'astronomie, et qui avait si bien profité des leçons d'un si grand maître. La petite ville de Perinaldo a donc eu l'avantage de produire trois astronomes célèbres; car on sait que Maraldi l'oncle et le premier Cassini étaient nés dans ce même lieu. Après avoir achevé ses études au collége des jésuites de San-Remo, le jeune Maraldi revint dans sa famille. Il ne fut pas long-tems incertain sur le choix d'un état; car son oncle ayant écrit pour proposer à son père de le lui envoyer, le jeune homme saisit avec joie l'occasion de venir en France, de voir cette belle contrée qui retentissait encore du nom de Louis XIV, et ce Paris dont il avait fait la capitale du monde et la patrie adoptive des savans.

Maraldi partit de Perinaldo au printems de l'année 1727; il n'avait encore que dix-huit ans. De combien de senti mens son ame dut être agitée en arrivant à Paris et en entrant à l'Observatoire! Présenté par son oncle au fils et au petit-fils de Jean-Dominique Cassini, accueilli par eux et par plusieurs autres savans réunis dans

G. Bernardi, *Giovanni Domenico Cassini*, Springer Biographies,
DOI 10.1007/978-3-319-63468-5

ce temple de l'astronomie, il dut se trouver beaucoup plus heureux que ces anciens philosophes de la Grèce que les prêtres d'Egypte n'admettaient aux mystères de l'initiation, qu'après de vives sollicitations et de longues épreuves. Le zèle, le dévouement et l'application que Maraldi montra dans ses premières études astronomiques, le rendirent bientôt digne de l'accueil qu'il avait reçu. Quelqu'agréable cependant que fût son noviciat, il ne fut pas tout-à-fait exempt de rigueurs. Une chambre de huit pieds carrés, pratiquée dans l'embrasure d'une fenêtre d'une des grandes salles de l'Observatoire, fut le premier appartement et le seul qu'il fût possibe de donner au nouvel astronome.

L'architecte qui avait tracé la distribution de l'Observatoire, n'y avait omis que les cheminées et les Iogemens. Il se proposait, dit-on, de loger les observateurs dans des bâtîmens extérieurs. En conséquence, l'édifice n'était composé que de grandes salles voûtées où l'on a eu par la suite beaucoup de peines à pratiquer un très-petit nombre d'appartemens qui pussent offrir les principales commodités, auxquelles un astronome ne renonce pas tout-à-fait, quoiqu'il passe plus de la moitié de sa vie dans le ciel. On ne doit donc par être étonné que le dernier arrivé fût aussi mal logé: mais il était fort éloigné de s'en plaindre; car cette petite cellule était très-conforme à son goût pour la solitude et à son caractère un peu sauvage, qui lui faisait trouver bon de ne pouvoir recevoir qu'une seule visite à la fois.

Au reste, c'était dans les vastes salles occupées par les instrumens qu'il fallait venir trouver le jeune Maraldi, si on voulait le voir; c'était là qu'il passait la moitié du jour et la plus grande partie de la nuit. Il s'y exerçait sans cesse au maniement des lunettes, à la pratique des observations, à la connaissance des étoiles j aussi fut-il bientôt en état de seconder son oncle dans l'entreprise qu'il avait formée de dresser un nouveau catalogue. Mais cet oncle, ce protecteur, ce maître si précieux vint à lui manquer au moment où il devait le moins s'y attendre.

Philippe Maraldi mourut deux ans après l'arrivée de son neveu. Cette perte eût été irréparable pour celui-ci, sans cette providence qui sait nous ménager des ressources dans le malheur, et qui lui fit trouver dans Jacques Cassini plus encore qu'il n'avait perdu; car ce bon parent l'adopta dès-lors au nombre de ses enfans, le constitua le frère et l'émule de son fils Cassini de Thury, un peu plus jeune que Maraldi, et qui se disposait à courir la même carrière. On verra dans la suite les deux cousins s'associer pour les mêmes travaux astronomiques et géographiques, et devenus presqu'en même tems membres de l'Académie, présenter à la fois trois noms de la même famille incrits sur la liste académique.

Les premières recherches de Maraldi se tournèrent vers la théorie des satellites de Jupiter, à laquelle il se consacra d'une manière particulière, et qui fut pendant cinquante ans son objet de prédilection, le but principal de ses observations. Ces petites planètes semblaient être un domaine de famille que Dom. Cassini avait acquis de Galilée, et qu'il avait transmis par héritage aux Maraldi ses neveux. En effet, on se rappelle quel accroissement de réputation s'acquit le premier Cassini lorsqu'en 1668, il publia les nouvelles Ephémérides des satellites de Jupiter. Plusieurs astronomes, et Galilée lui-même, avaient en vain tenté de calculer les mouvemens de ces petits astres; Cassini réussit le premier à determiner leurs révolutions, la durée de leurs éclipses, la grandeur de leurs orbites, la position de leurs noeuds. Ses tables, qu'il retoucha en 1693, furent long-tems les plus exactes.

Après lui Philippe Maraldi passa les vingt dernières années de sa vie à les perfectionner. En 1712 il avait fait la remarque importante que les durées des éclipses n'étaient pas toujours les mêmes a égale distance des noeuds, et avait découvert une variation dans l'inclinaison de l'orbite du quatrième satellite; la mort l'ayant surpris au milieu de ses recherches, son neveu Dominique Maraldi, dont nous faisons l'éloge, reprit le même travail, et y apportant autant de zèle et non moins de sagacité, il découvrit bientôt une semblable variation dans l'inclinaison de l'orbite du troisième satellite, et ayant en même tems reconnu une excentricité sensible dans l'orbite du quatrième, on lui fut redevable d'une nouvelle preuve de cette vérité, que les mêmes lois qui régissent notre système, gouvernent également le monde des satellites de Jupiter. C'est en 1732 que M. Maraldi par ces recherches intéressantes justifia l'adoption dont l'Académie l'avait honoré l'année précédente.

Son assiduité, sa persévérance pendant une longue carrière, à suivre les satellites de Jupiter, à les observer dans tous les points de leurs orbites et dans les circonstances les plus favorables, lui valurent encore par la suite d'autres découvertes. En 1765, il reconnut un mouvement d'oscillation dans le noeud du second satellite: et en 1769, il détermina la période des variations de l'inclinaison du troisième, qu'il trouva de 132 ans. Nous pouvons assurer, d'après le témoignage des registres de l'Observatoire, que depuis 1730 jusqu'en 1770, il n'a échappé à M. Maraldi d'éclipsés de satellites, que celles que le mauvais tems ou une absence forcée lui ont fait manquer. Il ne les a pas observées moins assidûment à Perinàldo, depuis 1770 jusqu'en 1785, où ses infirmités ne lui permirent plus

aucun genre d'occupation. Il faut en convenir; de toutes les observations astronomiques, celles des éclipses de satellites sont les plus pénibles à suivre. Elles n'ont lieu que la nuit, à des heures toutes différentes; elles sont fréquentes, et fatigantes pour la vue.

Mais, observer pendant les nuits, calculer pendant le jour, passer alternativement et pour se délasser de la contention de l'esprit aux fatigues du corps; telle est la vie de l'astronome. Il doit, de plus, ne se rebuter jamais ni des veilles, ni des préparatifs perdus, ni des voyages, ni des dangers, ni des sacrifices qu'un léger nuage, un simple brouillard peuvent rendre inutiles; enfin, il faut qu'à de vrais talens il réunisse la force, le courage et la patience. Voilà sans doute ce qui doit faire distinguer l'astronome des autres savans; voilà ce qui doit rendre plus rares et plus précieux ceux quijjnt les qualités nécessaires pour se livrer à l'astronomie. M. Maraldi les avait toutes, et il eut l'occasion d'en faire usage dans un autre genre de travail qui n'en exige pas moins.

Pendant huit années consécutives, de 1732 à 1740, il fut associé à son cousin Cassini de Thury, dans la description trigonométrique des côtes et des frontières de la France, ainsi que dans le tracé de ces méridiens, de ces parallèles et de ces perpendiculaires qui traversèrent le royaume dans tous les sens, et qui, liés ensemble par une chaîne continue de quatre cents triangles, appuyés sur dix-huit bases, formèrent le cannevas de la grande carte générale de la France, en 180 feuilles, qui a été publiée depuis. Cette carte, le plus grand monument élevé à la géographie, et le modèle de tous les travaux de ce genre, dont l'entreprise hardie a été poursuivie pendant près de cinquante ans au milieu des difficultés et des contrariétés, a dû son entière exécution au zèle opiniâtre de son auteur; et plus encore, à la générosité d'une société de citoyens dont les sacrifices patriotiques eussent été très-vantés, s'ils eussent eu lieu chez nos voisins; mais ils ont été longtems méconnus chez nous, et n'en méritent que davantage aujourd'hui la reconnaisance publique.

La feuille des triangles comprenant ces travaux fondamentaux de Maraldi et de Cassini de Thury, parut en 1744. Il serait assez naturel de penser que les recherches sur les satellites et les travaux géodésiques de notre académicien, étaient plus que suffisans pour nourrir son activité et employer tous ses momens: mais son goût pour le travail, son zèle pour le service de l'Académie, lui firent accepter un surcroit d'occupation, dont beaucoup d'autres auraient cherché à s'exempter avec d'aussi bons prétextes que ceux qu'il pouvait alléguer. En 1735, M. Maraldi fut chargé de la Connaissance des tems; cet ouvrage, que l'Académie royale des Sciences faisait publier tous les ans pour l'usage des astronomes, des marins et des voyageurs, était une tâche pénible et ingrate qu'elle imposait à celui de ses membres qui avait le courage de consacrer une grande partie de son tems à de longs calculs, auxquels ne sont attachés, il faut l'avouer, que bien peu de mérite et encore moins de gloire. Mais l'amour de la science, et le désir d'être utile, soutinrent M. Maraldi dans ce travail ingrat, pendant vingt-cinq ans, au bout desquels il fut remplacé par M. de Lalande.

Nous ne parlerons pas ici de tous les Mémoires de M. Maraldi insérés dans le Recueil de l'Académie des Sciences. Les principaux traitent de la théorie des satellites de Jupiter; nous avons indiqué les résultats importans que l'auteur y a établis. Mais nous déroberions quelque chose à sa gloire, si nous ne faisions remarquer que dans un Mémoire, lu en 1743, il donna le calcul de la comète de 1729 dans un orbite parabolique, et eut ainsi l'honneur d'être un des premiers en France à rendre hommage à la théorie newtonienne. Les astronomes français s'y décidèrent un peu tard: mais dans ce retard, il y eut sans doute plus de sagesse que d'esprit national. En effet, dans les sciences, c'est render hommage à la vérité de ne l'admettre qu'après un mûr examen. Ce qu'il y a de sûr, c'est que nos astronomes et nos géomètres sont ceux de tout le monde savant qui ont le plus généralement et le plus constamment rendu justice à la théorie de Newton, et que personne n'a plus contribué qu'eux à proclamer et à propager la gloire de son immortel auteur.

M. Maraldi que nous venons de présenter comme un savant laborieux, comme un astronome distingué, n'est pas moins intéressant à considérer sous un autre aspect. Doué d'une probité digne des premiers âges, il avait une austérité de moeurs et une franchise, qui n'étant point adoucies par des formes, pouvaient être prises quelquefois pour de la rudesse: mais cette rudesse n'était qu'une écorce, sons laquelle on trouvait bientôt la plus belle ame, le cceur le plus compatissant, le plus charitable. Il cherchait à la vérité à cacher aux autres et à lui-même sa sensibilité; mais elle perçait malgré lui, et les bonnes oeuvres qui en étaient le fruit, révélaient souvent son secret, non sans lui causer un vrai chagrin, je dirais même de l'humeur j car il n'aimait pas qu'elles fussent connues, encore moins qu'on lui en témoignât de la reconnaissance.

On le vit cependant se livrer sans réserve à cette sensibilité, et en donner les plus grandes marques dans deux occasions: à la mort de Jacques Cassini, son second père, et à celle de M. l'abbé de Lacaille, le seul homme avec lequel il eût formé une liaison intime. Le choix d'un tel ami, et la manière dont il sut apprécier deux hommes d'un si rare mérite, feraient seuls l'éloge de M. Maraldi. Accablé par ces deux pertes irréparables pour lui, il renonça presqu'à toute société, il ne sortait de l'Observatoire que pour aller à l'Académie, et ne vivait plus qu'avec ses livres, ses intrumens, et les satellites de Jupiter. La forte santé dont M. Maraldi avait toujours joui fut infiniment ébranlée vers la fin de 1763, par une enteroépiplocèle qui se déclara, mais dont les signes furent d'abord très-équivoques. Il fallut tout le talent de M. Tenon notre confrère pour ne pas s'y méprendre, et sur-tout pour réussir dans une opération délicate à laquelle il fut indispen sable d'en venir. Le traitement qui s'ensuivit dura trois mois entiers: le courage et la patience du malade furent inébranlables pendant une si longue épreuve. Quiconque connaît l'enthousiasme de M. Tenon pour son art, et son attachement pour tout ce qui tenait à l'ancienne Académie des Sciences, pourra juger du bonheur qu'éprouva ce savant anatomiste lorsqu'il put se dire: «J'ai fait une belle opération, et j'ai sauvé mon confrère.» Ajoutons que ce n'est pas la seule occasion où M. Tenon ait pu se rendre cette justice, et où il ait éprouvé une si douce jouissance.

Une grande maladie, dans la première jeunesse, ne produit communément que l'effet de ces orages au fort de l'été, qui passent et ne font que purifier l'air: mais dans un âge avancé, l'ébranlement qu'elle cause à des ressorts qui commencent à s'user, se prolonge jusqu'à la fin de la vie, et en rapproche souvent le terme. Quoique parfaitement guéri de l'accident qu'il avait éprouvé en 1763, M. Maraldi avait conservé de sa longue maladie une affection scorbutique, qui dérangeait fréquemment cette santé robuste dont il jouissait autrefois. Un de ses neveux, qui professait la médecine, vint en France passer quelque tems auprès de lui. Il lui persuada que le changement de lieu, de vie, et sur-tout l'air natal, pouvaient seuls le rétablir dans son ancien état. Il le pressa vivement de revenir au sein de sa famille recouvrer la santé. Comment ne pas se livrer à un si doux espoir? comment fermer l'oreille à la voix d'une famille qui nous rappelle, et le coeur à cet amour de la patrie que rien ne peut entièrement éteindre et qu'un rien sait rallumer?

M. Maraldi se décida donc à retourner à Perinaldo. Il lui en coûta sans doute de s'éloigner de l'Académie, et de sortir de l'Observatoire; mais depuis long-tems il n'observait plus que les éclipses des satellites de Jupiter, auxquels il avait voué un culte particulier; or, pour ce genre d'observations, Perinaldo lui offrait un ciel bien plus propice que celui de Paris. Il partit au mois d'avril 1770, emportant avec lui pendule, quart de cercle et lunettes, dont une de Campani, avec laquelle il observait depuis quarante ans. C'est avec cet attirail qu'il alla s'établir à Perinaldo. Cette ville qui avait vu naître plusieurs astronomes, méritait bien de posséder enfin un observatoire. Pendant quinze ans, M. Maraldi y poursuivit le cours de ses observations sur les satellites, et il ne manqua pas d'en faire part, de tems en tems, à l'Académie, à laquelle il resta constamment attaché de coeur et d'esprit jusqu'à son dernier soupir. C'est, au reste, ce que l'on peut dire de tous ceux qui ont eu le bonheur d'appartenir à ce corps illustre, et qui sont morts dans son sein.

L'air natal procura pendant plusieurs années à M. Maraldi ce qu'il s'en était promis: mais rien ne peut préserver des ravages dutèms, des infirmités de l'âge et des maux attachés à l'humanité. En 1785, notre académicien, âgé de soixante-seize ans, fut attaqué d'une espèce de maladie noire, fort singulière. Les accès se renouvelaient et disparaissaient tous les trois mois. Le malade n'y succomba qu'au bout de trois ans. Il mourut le 14 novembre 1788, avec une résignation et un courage que ne manque jamais d'inspirer la vraie philosophie, c'est-à-dire, la religion; car M. Maraldi fut du nombre, plus grand qu'on ne le pense communément, des savans qui se sont fait honneur de professer les principes du christianisme et d'en remplir les devoirs. Sa mort n'a point laissé de place vacante à l'Académie; il avait obtenu la vétérance deux ans après son départ. C'est M. Mechain qui l'a remplacé à l'Observatoire royal de Paris.

Appendix E
Tableau Chronologique de la vie e des ouvrages de J.D. Cassini

(Chronological Table of the Life and Works of J.D. Cassini)

1625.—Sa naissance à Périnaldo, le 8 juin.
1639.—Il entre au collége de Saint-Jérôme, tenu par les Jésuites à Gênes.
1646.—Il compose plusieurs pièces en vers latins et italiens, don't quelques-unes ont été imprimées (*Giustiniani gli scrittori liguri*, p. 571).
1649.—Il se rend à Bologne.
1650.—Le Sénat de Bologne lui donne la chaire d'astronomie, vacante par la mort de Cavalleri.
1652.—Il observe la comète qui paraît à la fin de cette année, et qui lui donne lieu de publier l'ouvrage suivant: *Mutinae, in-folio. Ad seren. princip. Franciscum Etenensem Mutinae ducem., Joan.-Domin. Cassini genuensis, in bononiensi archigymnasio public. Astronom. profess. de cometa ann.* 1652 et 1653.
Il suit dans ce premier ouvrage la fausse opinion de la formation des comètes par les exhalaisons de la terre, et même des étoiles qu'il suppose avoir un atmosphère, et démontre par ses observations que la comète est fort au-dessus de la région de la lune: mais depuis l'impression de cet ouvrage, la suite de ses bservations réforme ses idées; il s'aperçoit que la route de la comète peut être représentée par un mouvement régulier, qu'on peut en dresser des éphémérides et que l'astre se trouve fort au-dessus de Saturne. Il en écrit à Bouillaud, et se propose d'en publier incessamment une seconde partie sous le titre de *Theoria motus cometae anni 1652*: mais cet ouvrage annoncé n'a pas été imprimé (*Giustiniani, gli scrittori liguri*, p. 360). Il résout le problême déjà tenté par Kepler et jugé insoluble par lui et par Bouillaud, consistant à trouver directement et géométri quement l'apogée et l'excentricité d'une planète (*Opera Gassendi*, tome VI; *Hist. acad., Duhamel*, p. 55). Il n'a publié cette solution qu'en 1669 (*Histoire de l'Académie*, tome Ier[er], p. 111, et *Mémoires de l'Académie*, tome X, p. 488). Il écrit à Gassendi pour lui procurer une suite d'observations des planètes sur lesquelles il puisse établir les fondemens d'une nouvelle théorie des planètes (*Epistola Gassendi. Giustiniani*, p. 360).

G. Bernardi, *Giovanni Domenico Cassini*, Springer Biographies,
DOI 10.1007/978-3-319-63468-5

1654.—*Bononia, in-4°. Illustrissimis et sapientissimis senaloribus augustissimoe D. Petronii fabricae proefectis, novum lumen astronomicum ex novo heliometro.*
Bonon., in-folio. Controversice astronomicoe ad maximun heliometrum Petronii examini expositoe.
De novo gnomone meridiano in D. Petronii templo construendo: ad S. March. Innoc. Facchinetti confalon. (Manusc.)
1655.—Il trace la méridienne de Sainte-Pétrone, et publie une invitation aux mathématiciens de Bologne pour être témoins des operations faites à cette méridienne, et pour observer le solstice d'été de 1655. Il présente et dédie à la reine de Suède la *Description et l'usage de la méridienne de Sainte-Pétrone*, imprimée eu forme de thèse.
1656.—*Bonon., in-folio. Spécimen Obscrvationum Bononiensium, quoe novissimè in div. Petronii templo ad astronomioe novae constitutionem haberi caepere.*
1657.—Il est député à Rome avec le marquis Nicole- Tanari, par le Sénat, au sujet des différends élevés entre Bologne et Fcrrare sur le cours des eaux du Pô et du Reno. Il compose à ce sujet divers écrits, dont plusieurs sont imprimés, pour soutenir les intérêts de la ville de Bologne. (*Raccolta: del corso antico del Po e dei fiumi inferiori suoi tributarii.*)
Rom., in-4°. Alla Santità di nostro sign. Papa Alessandro settimo, per la sacra congregazione dell' acque, il regimento di Bologna.
Rom., in-4°. Idronomia nuova.
1659.—Il présente au Pape un planisphère gravé, avec ce titre: *Systema revolutionum superiorum planetarum circà terram ab anno 1659 ad sequentes per tricenos dies.*
Bonon., in-folio. Varie figure intagliate in rame, che rapresentano la prospettiva de pianeti con le' proporzione de' loro distanze al sole ed alla terra, periodiche revoluzioni, direzzioni, e retrogradazioni.
1661.—Il observe devant le duc de Modène l'éclipse de soleil de cette année, et à cette occasion il imagine la méthode de déterminer les longitudes terrestres par l'observation des éclipses de soleil, et celle de *tracer sur une carte géographique les apparences d'une éclipse de soleil pour tous les divers lieux de la terre.* Mais l'inquisiteur de Modène ne permit pas de publier celle-ci lorsque par la suite il voulut l'exposer dans un ouvrage intitulé: *Nova eclipsium methodus.*
1662.—*Mutinae in-folio. Novissimae motuum solis ephemerides ex recentioribus tabulis clariss. viri Joan.-Dom. Cassini in archigymn. Bonon. Astron. Profess. a March. Malvazia supputatoes cum epistolis auctoris ad Cassinum, ejusdemque responsis.*
1663.—Le Pape Alexandre VII lui donne la surintendance des fortifications du fort Urbin.
Bonon., in-folio. Joan.-Dom, Cassini epistola de observationibus in D. Petronii templo habitis.

1664.—Il est appelé à Rome par le Pape, et chargé des négociations pour régler avec la Toscane le cours de la Chiane; il observe avec la reine de Suède la comète de 1664, dont il ose, d'après les deux premières observations, tracer la route sur un globe, annoncer l'époque de sa plus grande proximité, prédire sa station et sa rétrogradation future, prédictions qui ont été justifiées. Le 3o juillet, se trouvant en Toscane à Città-della-Piave, il fait la *découverte des ombres des satellites sur le disque de Jupiter*. Plusieurs habites astronomes n'y veulent point croire, et lorsqu'ils les voyent, ils veulent les confondre avec les taches de Jupiter.

Ferrar., in-folio. Osservazione dell' eclisse solare fatta in Ferrara l'anno 1664, con una figura intagliata in rame, che rapresenta uno nuovo methodo di trovar l'apparenze varie che fu nel medesimo tempo in tutta la terra.

1665.—Une nouvelle comète paraissant au mois d'avril, il publie au bout de dix jours une table où la comète était calculée comme l'aurait pu être une ancienne planète.

Il détermine la révolution de Jupiter autour de son axe, et publie un grand nombre d'ouvrages, entre autres des éphémérides du mouvement des ombres des satellites sur le disque de Jupiter, pour répondre à ceux qui les confondaient avec les taches, et écrit plusieurs lettres à ce sujet, qui lui attirent de grandes discussions.

Rom. In-folio. Astronomicae epislolae duoe (8 augusti et pridie id. sept. 1665). Altera adm. Rev. P. Egidii Franc. Gottigniez. Soc. Jes. in Romano colleg. Profess. ad perillustr. et excell. D. Joan.-Domin. Cassinum, Bon. Arch. Astr. Altera ejusdem D. Cassini responsiva de difficultatibus circa éclipses in Jove a mediceis planetis affectas; aliaque noviter detecta, cum earum solutionibus. (Journal des Savans, 99–59.)

Bologna, in-folio. Due lettere astronomiche (gli 9 e 20 maggio) al sign. Abbat. Falconicri, soprà il confronto di alcuni osservazioni delle cometa di quest' anno 1665.

Roma, in-folio. Lettera astronomica (22 luglio 1665) al sign. Abb. Ottavio FalconII, soprà l'ombre de pianetini medicei in Giove, con le tavole dell' osservazioni ne due mesi seguenti, (Journ. sav., 22 fêvr. 1666.)

Rom., in-folio. Lettere astronomiche al sign. Abb. Ott. FalconII j soprà le varietà delle macchie osservate in Giove, e loro diurne rivoluzioni. (gli 12, 20 e 26 ottobre 1665.)

Rom., in-folio. Tabulae quotidianoe revolulionis macularum Jovis, nuperrime adinventoe a Joan.-Dom. Cassino.

Rom., in-fol. Theoria motus cometae anni 1664 et 1665, Pars prima, etc. (Miscellanea italiana, 1692.)

Rom., in-folio. Dissertatio apologetica de umbris mediceorum in Jove. con altre opere.

1666.—Il détermine la révolution de Mars autour de son axe, et réfute les objections de Riccioli, dans son Astronomie réformée, contre sa théorie des réfractions,

Bonon., in-folio. Martis circa proprium axem revolubilis, observations Bononienses. (Journ. sav., 25g-i5'j.)
Bonon., in-folio. Disceptatio apologetica de maculis Jovis et; Martis (Giustiniani, p. 371), en réponse à M. Serra.
Bonon., in-folio. De solis hypothesibus et de refractionibus. Syderum ad dubia admodum R. P. Jo.-Bapt. Riccioli, societatis Jesu, epistola ad Gemin. Montanari. (Miscell. ital. Et Journ. sav., 1693, 343–270.)
Romae, in-folio. Joan.-Domin. Cassini Opera astronomica.
1667.—Il aperçoit une *nouvelle étoile dans l'Eridan*, et détermine la révolution de Vénus autour de son axe. Il répond aux mauvaises difficutés de Levera sur la méridienne de Bologne; fait répéter le premier, en Italie, les expériences de la transfusion du sang, et en écrit une longue lettre au sénateur Berlingerigessi, qui l'a fait imprimer parmi ses autres ouvrages.
Estratto di una lettera al sign Petit, intorno a moto di Venere (Giorn. de litteral., 1668, p. 4°, et Mém. acad. t. X, p. 467).
Nuntii syderei interpres, et de planetarum facie, maculis et revolutione. (Ouvrage dont il n'y a eu que 64 pages d'imprimées, Journ. sav. y 182–123.)
Rom., in-folio. Dissertationes apologeticae de duplici gnomone in divi Petronii templo. (*Giustiniani, gli scritt. liguri*, p. 365.)
1668.—II est député par le Pape auprès du grand-duc de Toscane pour régler les confins de l'Etat ecclésiastique et de la Toscane. Il dédie au cardinal Caraffa une *Description des réjouissances solennelles qui ont eu lieu à Bologne*, au sujet de l'exaltation de Clément IX. Il apprend au mois de mai que le Roi de France désire qu'il soit en correspondance avec les membres de l'Académie royale des Sciences de Paris, et même que S. M. projette de le faire venir en France. Le Pape consent à ce voyage pour quelques années. En attendant, il publie une ébauche des *premières tables de satellites de Jupiter*, qu'il perfectionna par la suite en 1693. (*Hist. acad.*, tome Ier^{er}, p. 313, et *Mém. acad.*, tome VIII, p. 318). Mais plusieurs autres ouvrages qu'il avait commencé à faire imprimer se trouvent interrompus par son départ pour la France.
Bonon., in-folio. Ephemerides Bononienses mediceorum syderum, ex hypothesibus et tabulis Joan.-Domin. Cassini (Giorn. de litterat. 1668, et Journ. sav., 154–105.)
Bolog., in-folio. Spina celeste meteora, osservata in Bologna, il mese di marso 1668, da Giov. Domen. Cassini astron. Dello studio publico.
Bologn., in-4°. Apparizioni celesti dell' anno 1668 osservate in Bologna da Giov. Domen. Cassini. (Journ. sav., 57–42.)
Observations sur les insectes qui s'engendrent dans le chêne. (*Journ. sav.*, 100–71.)
Bonon., in-4°. Joan. Domin. Cassini Disceptatio apologetica, de maculis Jovis et Martis, annis 1666 et 1667; et de conversione Veneris circà axem.
Novite osservata nelle stelle fisse. (Giornal. de' litterat., 1668, p. 122.)
Osservazione fatte in Roma della Stella rinascente nel collo della Balena li 14 gennaro, per aviso ricevuto dal sign. Cassini. (*Hist. acad.*, tome Ier, p. 132; *Giorn. de' litter.*, pp. 7–36.)

Osservazione dell' eclisse solare fatta in Bologna li 24 nov. 1668. (Giorn. de' litter., 1668, pp. 154)
Osservazione di una eclîssa lunare fatta in Roma la notte seguente lo 25 maggio 1668. (Giorn. de' litter., 1668, p. 71.)
Observations de la comète de 1668. (Acta. Reg. Soc. Lond.)
1668.—Astronomia geometrica. (Lettera al Gassendo.)
Geodesia nova. (Geographia reformata. Riccioli.)
Astronomia optica (lettere al Malvazia e al Montanari.)
Plantographia nuova.
Almagestum promotum. Ces cinq derniers ouvrages n'ont pas été achevés ni publiés.
1669.—Il part de Bologne le 25 février pour se rendre en France, arrive le 4 avril à Paris, et est présenté au Roi le 6.
Bonon. in-4°. Nova ratio inveniendi metricè et directe apogoea, excentricitates et anomalias motus planetarum (Hist. acad., tome Ier, p. 110; Mém., tome X, p. 488; Journ. sav., 34–22.)
1670.—Observations sur l'étoile nouvellement découverte proche de la tête du Cygne, par D. Anthelme, chartreux de Dijon. Table des changemens et apparitions de la changeante du col de la Baleine. (Hist. acad., tome Ier, p. 132; Mém., tome X, p. 496; Journ. Sav.)
Méthode pour trouver la différence des longitudes des lieux par les observations correspondantes des phases des éclipses de soleil. (Hist. acad., tome Ier, p. 133.)
1671.—Il s'établit à l'Observatoire royal le 14 septembre; observe en mai, août et novembre les phases de l'anneau de Saturne, et découvre près de cette planète un nouveau satellite, qu'il aperçoit pour la première fois vers la fin d'octobre et au commencement de novembre, mais qu'il perd bientôt de vue. (Hist. acad., tonie I", p. 115, 150 et 174.)
Il aperçoit cinq nouvelles étoiles dans Cassiopée. (Hist. acad., tome I^{er,} p. 224.)
Observations sur une nouvelle étoile qui tellation du Cygne, et sur les étoiles qui paru proche de la constellation du Cygne, et sur les étoiles qui paraissent et disparaissent de tems en tems. (Journ. sav., pp. 34–61.)
Observations des satellites de Jupiter, 'correspondantes à celles de M. Picard à Uranibourg. (Mém. acad., tome VII, p. 228.)
Paris, in-4°. Nouvelles observations des taches du soleil, faites à l'Académie royale des Sciences, les 1 1, 12 et 13 août 1671.
1672.—Il s'occupe particulièrement à Paris des observations correspondantes à celles que doit faire M. Richer à Cayenne, ainsi qu'à déterminer, par l'ingénieuse méthode qu'il avait imaginée, la parallaxe de Mars et celle du soleil qu'il réduit à 9 secondes et demie ou 10 secondes. Il va faire des observations en diverses provinces pour la perfection de la géographie du royaume. Il découvre encore un nouveau satellite à Saturne, en recherchant celui qui lui avait échappé ennovembre 1671, et qu'il revoit en même tems. (Hist. acad., tome Ier, p. 159.) Le 25 janvier 1672, il croit apercevoir un satellite à Vénus. (Mém. acad., tome VIII, p. 183.) Il propose son idée d'un zodiaque pour les comètes (Hist. acad., tome Ier,

pag,160.) Histoire de la découverte de deux planètes autour de Saturne, faite en 1671 et 1672. (Mém. acad., tome X, p. 584; Journ. sav., 1677, pp. 70–40)
Observations de la comètes autour de 1672 avec des réflexions. (Mém. Acad., tome Ier, p. 160; tome X, p. 518 et 526; Journ. Sav., 73–29 et 84–35.)
Relation du retour de la grande tache permanente dans Jupiter, vue en 1665 (Mém. Acad., tome X, p. 513; Journ. Sav., 68–25), et qui a servi a déterminer la révolution de cette planète autour de son axe.
Observations faites en diverses endroits du royaume. (Mém. Acad., tome VII, p. 349; Recueil d'Observ., in-folio, 1693.)
1673.—Il est naturalisé Français et s'établit.
Il revoit en février le premier satellite de Saturne qu'il avait découvert en 1671, lequel a la singularité de disparaître lorsqu'il est dans la partie orientale de son cercle la plus éloignée de Saturne.
Paris, in-folio. Découverte de deux nouveaux satellites autour de Saturne. (C'est le 5e et le 3e qui depuis les découvertes de M. Herschell sont devenus le 7e et le 5e.)
1674.—Le retour de M. Richer à la fin de 1673 confirme ses théories, l'historien de l'Académie s'exprime ainsi: On eût dit que M. Cassini s'était entendu avec les astres. Ce qu'il avait conjecturé devint indubitable, et ses suppositions se changèrent en principes; le ciel décida absolument pour les réfractions et les parallaxes de M. Cassini. (Hist. acad., tome Ier, p. 271.)
Les élémens de l'astronomie vérifiés, par le rapport des tables de M. Cassini aux observations de M. Richer à Cayenne. (Mém. acad., tome VIII, p. 55; Recueil d'Observ., in-folio, 1693.)
1675.—Observation de l'éclipsé de soleil du mois de janvier. (Journ. sav.)
Pleni lunii ecliptici die 7 jul. Schema.
Hypotheses circà motus librationis lunae.
Observations des éclipses du 11 janv. et 7 juillet. (Hist. acad., tome Ier, p. 205; Mém., tome X, p. 544 et 555.)
Observations nouvelles touchant le globe et le double anneau de Saturne. (Mém. acad., tome X, p. 582; Journ. sav., 1677.)
1676.—Extrait d'une lettre à l'auteur du Journal des Savans, contenant quelques avertissemens aux astronomes touchant les configurations des satellites de Jupiter pour les années 1676 et 1677 (Hist. acad., t. Ier, p. 212; Mém., t. X, p. 572; Journ. sav., 214–122.)
Paris, in-4°. La méthode de déterminer les longitudes des lieux de la terre par les observations des satellites de Jupiter, vérifiée et expliquée par M. Cassini. (Mém. acad., tome X, p. 569; Journ. sav., 192–109.)
Observations de l'éclipsé du soleil du 11 juin, faites en plusieurs endroits de l'Europe. (Mém. acad., tome X, p. 571.)
Description du mouvement qu'a fait une tache dans le soleil sur la fin de novembre 1676. (Hist. acad., tome Ier, p. 216; Mém. acad., tome X, p. 578; Journ. sav., 239–135.)
Relation d'un feu prodigieux qui parut à Rome, à Bologne et autres lieux d'Italie, le 31 mars 1676. (Journ. sav., 1 18–66.)

Balance arithmétique, sa description et son usage pour connaître les nombres par les poids (Hist. acad., tome Ier, p. 217; Machines acad., tome Ier, p. 143; Journ. sav., 253–145.)
1677.—Il aperçoit pour la première fois sur Saturne une bande parallèle à la ligne des anses, traversant le disque très-près du centre. (Hist. acad., tome Ier, p. 376; Journ. sav., 56–32.) Il propose un nouveau jovilabe ou instrument pour représenter les mouvemens et les configurations des satellites de Jupiter. (Hist. acad., tome Ier, p. 248; Explicatio jovilabii Cassiniani, Weidler, 1727, Wittemb.) Il donne, dans une seule proposition très-simple et très-ingénieuse, la théorie de la projection des bombes (Hist. acad., tome Ier, p. 255,) Les places qu'il possédait en Italie sont supprimées.
Paris, in-4°- Nouvelle théorie de la lune. (Mém. acad., tome X, p. 589; Journ. sav., 117–66.)
Réflexions sur les observations de Mercure dans le soleil, vu à la Chine. (Mém. acad., tome X, p. 599; Journ. sav., 244–159.)
Observations sur la rotation de Jupiter et autres changemens dans cette planète. (Lectures and Collecl. Rob. Hooke.)
Tache sur Jupiter aperçue le 5 juillet 1677. Vérification de la période de la révolution de Jupiter autour de son axe par des observations nouvelles. (Mém. acad., tome X, p. 526; Journ. sav., 314–134.)
Avis sur la comète de 1677. (Mém. acad., tome X, p. 582.)
Théorie de la comète qui a paru aux mois d'avril et de mai 1677, tirée des observations des plus célèbres astronomes. (Hist. acad., tome Ier, p. 236; Mém. acad., tome X, p. 592; Journ. sav., 120–68.)
Suite des observations faites à l'Observatoire royal, touchant la tache qui a paru dans le soleil, les mois décembre 1677. (Mém. acad., tome X, p. 581; Journ. sav., 8–6.)
Apparences météorologiques, observées à Paris le 17 mai 1677, d'une croix blanche autour de la lune, d'une couronne autour du soleil, et de trois faux soleils. (Mém. acad., tome X, p. 583.)
Avis aux astronomes sur le retour de l'étoile de la Baleine. (Hist. acad., tome Ier, p. 238; Mém. acad., tome X, p. 600.)
1678.—Il observe au mois de mai Vénus à moitié illuminée comme la lune dans son premier quartier. Les équinoxes de cette année, déterminées avec soin, se trouvent conformes à ses tables du soleil, mais elles sont éloignées de trois heures des tables rudolphines. Il aperçoit des taches sur Jupiter dans l'endroit où doivent se trouver les satellites dans leur conjonction inférieure, et il en conclut que les satellites ont des taches, qu'ils nous paraissent plus petits qu'ils ne sont en effet, et qu'ils ont un mouvement sur leur axe. Il soupçonne un atmosphère au premier satellite. (Hist. acad., t. Ier, p. 266.)
Observation d'une nouvelle tache dans le soleil. (Mém. acad., tome X, p. 601; Journ. sav., 88–49.)
Occultation de Saturne par la lune le 27 février. (Hist. acad., tome Ier, p. 264; Mém. acad., tome X, p. 602.)

Observations de taches et facules dans le soleil au mois de mai. (Mém. acad., tome X, p. 604; Journ. sav., 248–132.)
Observation de l'éclipse de lune du 29 octobre. (Hist. acad., tome Ier, p. 264; Mém. acad., tome X, p. 612.)
Observation d'une étoile double au front du Scorpion. (Hist. acad., tome Ier, p. 266.)
1679.—Réglement des tems par une méthode facile et nouvelle par laquelle on fixe pour toujours les équinoxes au même jour de l'année. (Hist. acad., tome Ier, p. 514; Mém. acad., tome X, p. 615; Journ. sav., 97–55 et 113–64.)
Méthode de rétablir l'usage du nombre d'or pour régler toujours les épacles d'une même façon. (Mém. acad., tome X, p. 618.)
Observation de l'éclipse de Jupiter et de ses satellites par la lune, le 5 mai. (Hist. acad., tome Ier, p. 303; Mém. acad., tome X, p. 620; Journ. sav., 191–105.)
Découverte d'une tache extraordinaire dans Jupiter le 29 mai. (Journ. sav., 1686.)
1680.—Il imagine une Nouvelle progression de norneres applicable à la théorie des planètes, et qui offre plusieurs belles propriétés. (Hist. acad., tome Ier, p. 309.) Il revoit le 8 avril la grande tache de Jupiter (Hist. acad., tome Ier, p. 514.), et corrige ses premières tables des satellites de Jupiter, et particulièrement celles du premier dont il prend, pour nouvelle époque du mouvement, l'immersion du 21 juillet 1680
à 13 heures 54 minutes. (Hist. acad., tome Ier, p. 313.)
Ephémérides des satellites de Jupiter pour les années 1681 et 1683. (Hist. acad., tome Ier, p. 331.)
Taches sur le disque du soleil, observées le 20 mai. (Hist. acad., tome Ier, p. 317.)
1681.—N'ayant encore observé qu'une fois la comète du mois de décembre 1680, il prédit au Roi, en présence de toute la cour, qu'elle suivra la même route que la comète observée par Tycho en 1577. Elle la suivit en effet.
Paris, in-4°. Observations et réflexions sur la comète qui a paru aux mois de décembre 1680 et de janvier 1681, présentées au Roi. (Journ. sav., 145–96.)
Abrégé des observations sur la comète de 1680. (Hist, acad., tome Ier, p. 331.)
Instruction générale pour les observations astronomiques et géographiques à faire dans les voyages (Mém. acad., tome VII, p. 432.)
Observation de Vénus dans le parallèle du soleil pour la détermination de sa parallaxe et de sa distance à la terre. (Hist. acad., tome Ier, p. 331.)
Nouveau planisphère d'argent fait et présenté au Roi; sa description et ses usages. (Hist. acad., tome Ier, p. 317, et tome II, p. 100; Journ. sav., 317–148; Mach. acad., tome Ier, p. 133.)
Observation de l'éclipsé de lune du 39 août 1681. (Hist. acad., tome Ier, p. 331; Acta Lips., 1682.)
1682.—Paris, in-4°. Premières observations de la comète de 1682, présentées au Roi. (Journ. sav., 209–188.)

Méthode pour trouver la parallaxe de Vénus par sa comparaison avec une étoile qui se rencontre dans le même parallèle que cette planète, invitation aux astronomes d'en faire usage en février 1683. (Hist. acad., tome Ier, p. 251.)
Réflexions sur deux éclipses de lune de cette année. (Hist. acad., tome Ier, p. 349.)
Réflexions sur l'éclipsé observée à Jutthia, le 22 février, par le père Thomas. (Mém. acad., tome VII, p. 694.)
Expériences sur un grain de phosphore sec. (Hist. acad., tome Ier, p. 343.)
1683.—Il découvre la lumière zodiacale le 18 mars, la même qu'il avait déjà aperçue à Bologne le 10 mars 1668. Il commence au midi de Paris les mêmes opérations que Picard avait faites vers le nord pour la mesure du degré du méridien. Il communique à l'Académie une méthode de déterminer la parallaxe des planètes, une théorie des étoiles fixes, une théorie de Vénus, et diverses additions et corrections au calendrier. (Hist. acad., tome Ier, p. 383.)
Nouveau phénomène rare et singulier d'une lumière céleste qui a paru au commencement du primeras. Comparaison de cette apparence à d'autres semblables. (Mém. acad., tome X, p. 637 et 640; Journ. sav., 119–76.) Découverte de la lumière céleste qui paraît dans le ciel. (Hist. acad., tome Ier, p. 378; Mém. Acad., tome VIII, p. 121; Recueil d'observations, in-folio, 1693.)
Réflexions sur les bandes et la rotation de Saturne. (Hist, acad., tome Ier, p. 376.)
Histoire de quelques parélies vus en avril et mai en différens endroits. (Mém. acad., tome X, p. 646; Journ. sav, 191–119.)
Réflexions sur les noeuds de la lune et sur son éclipse observée à Macao, en juillet, par le père Thomas. (Mém. acad., tome VII, p. 701.)
Observations sur une liqueur renfermée dans une bouteille et qui fume aussitôt qu'on ôte le bouchon.
1684.—Il découvre encore au mois de mars deux nouveaux satel lites très-proches de Saturne, parle moyen d'objectifs de Campani de 100 pieds et de 136 pieds de foyer. On frappe à cette occasion une médaille qui représente le système de Saturne avec cette légende: Saturni satellites primum cogniti. (Hist. acad., tome Ier, p. 415.) Saturne se trouvait alors accompagné de cinq satellites, dont le quatrième découvert par Huguens en 1655, les quatre autres, premier deuxième, troisième et cinquième, découverts par Cassini. (Depuis, en 1787 et 1789, M. Herschell en a découvert encore deux plus près que tous les autres du corps de Saturne.) Il achève la description du méridien depuis Paris jusqu'à Bourges, travail qui ne fut repris et poursuivi qu'en 1700.
Nouvelle découverte des deux satellites les plus proches de Saturne (Mém. acad., tome X, p. 584 et 694; Journ. sav., année, 1686, 97–77.) Ils sont devenus le troisième et le quatrième depuis les découvertes de Herschell.
Description d'une tache qui a paru dans le soleil au mois de mai facules observées à sa place au mois de juin, et retour de la tache à sa première forme. (Hist. acad., tome Ier, p. 408; Mém. acad.: tome X, pp. 653–661.)
Remarques sur la parallaxe de Mars périgée, détermination de celle du soleil. (Hist. acad., tome Ier, p. 418.)

Observation de l'éclipsé de lune du 37 juin et de l'éclipsé de soleil du 12 juillet. (Hist. acad., tome Ier, p. 411; Mém. acad., tome X, p. 664 et 667; Jour. sav., 309–197.)
Observation de l'éclipsé de lune du 21 décembre (Mém. Acad, tome X, p. 674.)
Réflexions sur l'observation de l'éclipsé de lune, faite à Goa par le père Noël. (Hist. acad., tome Ier, p. 436; Mém. acad., tome VII, p. 648.)
1685.—Observation de l'éclipsé de lune du 10 décembre 1685, avec la supputation dés longitudes de divers lieux, tant dans le royaume qne dans lès pays étrangers où elles ont été faites. (Hist. acad., tome Ier, p. 434; Mém. acad., tome X, p. 700: Journ. sav., 1686, 317–366.)
Manière d'employer des tuyaux pour les objectifs à longs foyers. (Hist. acad., tome Ier, p. 432.)
Remarques sur la grande et ancienne tache de Jupiter qui n'avait pas paru depuis six ans. (Hist. acad., tome Ier, p. 443)
Sur un poisson qui fait l'effet d'un baromètre. (Hist. acad., tome Ier, p. 424.)
Sur diverses pierres dures dans lesquelles on trouve de petits animaux bons à manger. (Hist. acad., tome Ier, p. 426.)
Sur la décharge du Reno dans le Pô, proche Ferrare. (Hist. acad., tome Ier, p. 443.)
Expérience sur la quantité d'eau nécessaire pour faire aller un moulin. (Hist. acad., tome Ier, p. 444.)
1686.—Lettre au révérend père Gouye sur les observations de l'éclipsé de Jupiter par la lune, faites à Paris et à Avignon le 10 avril. (Hist. acad., tome II, p. 11; Mém. acad., tome X, p. 704; Journ. sav., 167–132.)
Découverte d'une tache extraordinaire dans Jupiter. (Hist. acad., tome II, p. 11; Mém. àcad., tome X, p. 707; Journ. sav., 201–161)
Réflexions sur les observations des révérends pères Jésuites, faites à Lôuveau. (Mém. àcad., tome VII, p. 630.)
Sur la période de 600 ans et le retour des éclipses de lune au bout de 669 mois. (Hist. acad., tome II, p. 13.)
Sur les cinq satellites de Saturne; rectification de leur période et, de leurs mouvemens. (Hist. acad., tome II, p. 13.)
Observation de l'éclipsé de lune du 10 décembre. (Mém. acad., tome X, p. 709.)
Observation d'une tache sur le solel, le 4 mai.
1687.—Il croit revoir un satellite à Vénus, ainsi qu'en 1672. Il fait lecture à l'Académie de son Traité de l'origine et du progrès de l'Astronomie, qui n'a été imprimé qu'en 1693; d'un traité des éclipses de soleil; d'une méthode nouvelle d'observer les conjonctions des planètes, et d'une théorie de Jupiter. (Hist. acad., tome II, p. 32.)
Observation d'une météore en forme de globe de feu de la grandeur de la lune (Hist. acad., tome II, p. 32.)
Expériences sur les aimans de M. Petit et du père. Grandami, et conjeciures sur la cause du changement de déclinaison des aiguilles aimantées. (Hist. acad., tome II, p. 16 et 18.)

Observation sur une fontaine proche de Bologne qui prend feu en y approchant une chandelle. (Hist. acad., tome II, p. 23.)

Avis sur une grande comète vue en août 1686 à l'embouchure de la rivière des Amazones. (Hist. acad., tome II, p. 31.)

1688.—Il lit à l'Académie une dissertation sur le jour auquel on doit célébrer la fête de Pâques; répond aux objections de Vossius contre la méthode de déterminer les longitudes par les éclipses de satellites de Jupiter. Il va en Picardie et en Artois faire diverses observations pour la perfection de la géographie.

De la méthode de déterminer les longitudes des lieux de la terre par les observations des satellites de Jupiter. (Mém. acad., tome VII, p. 715; Journ. sav., deuxième Partie, 198–165.)

Lettres sur quelques corrections à faire à la théorie du cinquième satellite de Saturne. (Acta Lips., 1688.)

Observations sur différentes taches du soleil vues cette année, avec une méthode nouvelle pour déterminer la révolution du soleil. (Hist. acad., tome II, p. 57; Journ. sav., 167–139.)

1689.—Il fait connaître à l'Académie sa divination des règles de l'astronomie indienne, et lui présente un nouvel instrument pour prendre les verticaux. (Hist. acad., tome II, p. 74.)

Règles de l'astronomie indienne pour calculer les mouvemens du soleil et de la lune. Réflexions sur la chronologie chinoise, note sur l'île Taprobane. (Hist. acad., tome II, p. 71; Mém. acad., tome VIII, p. 214, 300, 312; Journ. sav., 1691, 182–138; Relation de Siam, Laloubère, tome II; Recueil d'observations, in-folio, 1693.)

Réflexions sur la longitude de la côte orientale de la Chine. (Mém. acad., tome VII, p. 793.)

1690.—Il corrige et refond en entier ses tables de satellites de Jupiter. Il rejette l'idée qu'il avait eue le premier du mouvement successif de la lumière. (Hist. acad., tome II, p. 109; Mém. acad., année 1707, p. 26 et 78.) Le roi d'Angleterre étant venu le 23 août visiter l'Observatoire, il célèbre cette visite dans une pièce de vers latins imprimée et qu'il lui présente.

Réflexions sur l'observation de Mercure dans le soleil, faite à la Chine par le père Fontenai en 1690, et publiée par le père Gouye. (Hist. acad., tome II, p. 194; Mém. acad., tome X, p. 308.)

Observations sur de nouvelles taches et de nouvelles bandes dans le disque de Jupiter. (Hist. acad., tome II, p. 104; Mém. acad., p. 104; tome X, p. 598.)

1691.—Paris, in-4°. Nouvelles découvertes dans le globe de Jupiter, faites à l'Observatoire royal. (Journ. sav., 51–38.)

Observation d'une conjonction précise d'un satellite de Saturne avec une étoile fixe, le 19 juin. (Hist. acad., tome II, p. 158; Mém. acad., tome X, p. 74; Journ. sav., 1692, 215–163.)

Projet pour la continuation de la méridienne dans toute l'étendue du royaume. (Hist. acad., tome II, p. 131.)

1692.—Il fait graver une grande figure de la lune dont, pendant un intervalle de sept années, il s'était occupé d'observer chaque tache l'une après l'autre, et d'en faire faire des dessins parfaitement exé cutés en grand par Patigny (1). [(1) J'ai fait voir à l'Académie des Sciences, en 1787, la collection précieuse de ces dessins recueillis dans un atlas format grand-aigle, et j'ai publié la réduction de cette grande figure de la lune, gravée par Jamnet, avec des observations très-curieuses qui n'avaient point été publiées.]

Nouvelles découvertes de diverses périodes de mouvement dans la planète de Jupiter depuis le mois de janvier 1691 jusqu'au commencement de l'année 1692. (Hist. acad., tome II, p. 130 et 158; Mém. acad., tome X, page Iere; Journ. sav., 84–64.)

Observation de la figure de la neige. (Hist. acad., tome II, p. 141; Mém. acad., tome X, p. 37; Journ. sav., 120–91.)

Remarques sur la longitude et la latitude de Marseille. (Hist. acad., tome Ier, p. 163; Mém. acad., tome X, p. 156.)

Observation d'un nouveau phénomène en forme de lance, faite à l'Observatoire royal le 20 mars. (Mém. acad., tome X, p. 90.)

Observation sur la conjonction de la lune et de Mars au mois d'avril. (Hist. acad., tome II p. 159; Mèm. acad., tome X, p. 98; Journ. sav., 215–163.)

Observation du passage de la planète de Mars par la nébuleuse de l'écrevisse au mois de mai. (Hist. acad., tome II, p. 159; Mèm. acad., tome X, p. 115.)

Observation faite en plein jour, le 19 mai, d'une éclipse de Vénus par la lune. (Hist. acad., tome II, p. 160; Mèm, acad., tome X, p. 138.)

Avertissement sur l'éclipse de lune qui doit arriver la nuit du 28 juillet, et observation de cette éclipse avec une méthode pour déterminer les longitudes par diverses observations d'une même éclipse, interrompues et faites en difiérens lieux. (Hist. acad., tome II, p. 160; Mèm. acad., tome X, p. 126 et 150; Journ. sav., 431–324–243.)

Eclipses des satellites de Jupiter qui auront lieu en 1693. (Mém. acad., tome X, p. 179.)

Observation de la conjonction de Vénus avec le soleil, du 2 septembre. (Hist. acad., tome II, p. 161, Mèm. acad., tome X, p. 198.)

De la révolution des quatre satellites autour de Jupiter, compare à celle de Jupiter autour du soleil. (Hist. acad., tome II, p. 158.)

Diverses observations sur Jupiter après sa conjonction avec le soleil, le 9 juin 1692.

1693.—Paris, in-folio. De l'origine et du progrès de l'astronomie, et de son usage dans la géographie et dans la navigation. (Recueil d'observations faites par ordre du Roi; Mém acad., tome VIII, page Iere; Journ. sav., 1697, 111–98, et 1727, 436.)

Réflexions sur l'observation faite à Marseille par M. de Chazelle, de l'éclipse de lune du 22 juin. (Hist. acad., tome II, p. 191; Mèm. acad., tome X, p. 240; Journ. sav., 132–99.)

S'il est arrivé du changement dans la hauteur du pôle ou dans le cours du soleil. (Hist. acad., tome II, p. 194; Mèm. acad., tome X, p. 360.)

Sur une apparition nouvelle de l'ancienne tache de Jupiter au mois de février. (Hist. acad., tome II, p. 193.)
Paris, in- 4°. Les hypothèses et les tables des satellites de Jupiter réformées sur de nouvelles observations. (Mèm. acad., tome VIII, p. 317.)
Observation sur une éclipse de soleil vue à Paris en juillet. (Hist. acad., tome II, p. 192.)
Observation de deux parasélènes et d'un arc-en-ciel dans le crépuscule, le 10 juin. (Hist. acad., tome II, p. 103; Mèm. acad., tome X, p. 400.)
Description de l'apparence de trois soleils vus en même tems sur l'horizon le 18 janvier. (Hist. acad., tome II, p. 167; Mèm. acad., tome X, p. 234; Journ. sav., 132–99.)
Dissertation sur l'état des eaux à Ferrare. De la largeur et de la profondeur du Pò à Lago Scuro. (Hist. acad., tome II, p. 173.)
1694.—Observation de l'éclipsé de lune du 7 juillet, et réflexions. (Hist. acad., tome II, p. 228.)
Observation des taches de Jupiter et de ses satellites; variations qu'elles peuvent causer à leurs éclipses. (Hist. acad., tome II, p. 224.)
Remarques sur le mouvement de l'étoile polaire en longitude et vers les pôles du monde. (Hist. acad., tome II, p. 229.)
Réflexions sur la conjonction de Mercure avec le soleil, dont les anciens et les modernes ont fait mention. (Hist. acad., tome II, p. 229.)
Relation d'une éruption du Vésuve. (Hist. acad., tome II, p. 204.)
Observation sur un baromètre lumineux. (Hist. acad., tome II, p. 202.)
Sur le mouvement d'oscillation de feuilles de papier suspendues par deux fils. (Hist. acad., tome II, p. 221.)
1695.—Il retourne en Italie et vérifie la méridienne de Sainte-Pétrone.
Bolog., in-folio. La meridiana del tempio di S. Petronio tirata e preparata per le osservazioni astronomiche l'anno 1658, revista e restaurata dal sign. Giov. Domen. Cassini. (Hist. acad., tome II, p. 265.)
Observation de l'éclipse de lune du 20 novembre, faite à Bologne. (Hist. acad., tome II, p. 264; Mém. acad., tome VII, p. 515.)
Observation de l'éclipse de soleil du 6 décembre, faite à Gènes. (Mém. acad., tome VII, p. 521.)
1696.—Observations astronomiques faites en France et en Italie, en 1694, 1695 et 1696. (Hist. acad., tome II, p. 268, 277, 292; Mém. acad., tome VII, p. 463.)
Observations faites proche le solstice d'hiver. (Hist. acad., tome II, p. 290.)
1697.—De la justesse admirable de la correction grégorienne des cycles lunaires. (Mém. acad., tome X, p. 739; Journ. sav., 80–71.)
Réflexions sur l'ancien canon pascal de Saint-Hippolyte. (Hist. acad., tome II, p. 300.)
Réflexions sur le calendrier et sur la différence entre les cycles lunaires et solaires. (Mém. acad., tome II, p. 318.)
Réflexions sur deux éclipses de l'année 1677 et principalement sur celle de lune employée à l'examen du calendrier. (Hist. acad., tome II, p. 322.)

Sur l'étoile changeante du col de la Baleine et sur la conjonction écliptique de Mercure et du soleil, observée le 3 novembre. (Hist. acad., tome II, p. 331.)
1698.—Réflexions sur les intervalles de tems entre les éclipses des satellites de Jupiter, comparés au retour de Jupiter à son aphélie. (Hist. acad., tome II, p. 343.)
Corrections à faire à la révolution et à la première équation du premier satellite de Jupiter. (Hist. acad., tome II, p. 344.)
Réflexions sur Mercure dans sa plus grande digression du soleil. (Hist. acad., tome II, p. 344.)
1699.—Observation de l'éclipse de lune du 15 mars. (Mém. acad., année 1699, p. 13.)
Du retour et du zodiaque des comètes. (Hist. acad., tome II, p. 342; Mém. acad., année 1699, p. 36.)
Observation de trois nouvelles taches sur Jupiter. (Mém. acad., année 1699, p. 103.)
Observation de l'éclipse de soleil du 23 septembre, et réflexions. (Mém. acad., année 1699, pp. 103–274.)
Comparaison des observations de la comète de 1699, faites à la Chine par le père Fontenai, rapportées avec celles qxii furent faites à l'Observatoire royal de Paris; description des quatre étoiles proche du cercle polaire avec lesquelles on commença d'apercevoir cette comète à Paris. (Mém. acad., année 1701, pp. 50–59.)
1700.—Il reprend à Bourges les opérations de la description de la partie méridionale du méridien qu'il avait commencée en 1683 et 1684 (Mém. acad. année 1700; Hist., p. 123.)
Réflexions sur les observations faites en Bothnie. (Mém. acad., p. 39.)
Nouvelles règles pour trouver les épactes des centièmes années non bissextiles. (Hist. acad., p. 110.)
1701.—De la méridienne de l'Observatoire royal, prolongée jusqu'aux Pyrénées. (Mém. acad., p. 171.)
Observation de l'éclipse de lune du 22 février, et comparaison des phases principales observées en différentes villes d'Europe. (Mém. acad., p. 65 et 68.)
Sur des taches dans le soleil, observées à Montpellier le 29 mars, (Mém. acad., p. 78.)
De la correction grégorienne des mois lunaires ecclésiastiques. (Mém. acad., p. 367.)
Avis sur la nouvelle réforme du calendrier. (Mém. acad., Hist., pp. 106–108.)
1702.—Il prolonge les opérations de la description du méridien jusqu'au mont Canigou.
Comparaison des mesures itinéraires anciennes avec les modernes. (Mém. acad., p. 15.)
Observation d'un nouveau phénomène, faite à l'Observatoire le 2 mars, avec quelques réflexions et diverses autres observations sur une même comète. (Mém. acad., p. 107.)
Observations d'une comète vue à Rome au mois d'avril. (Mém. acad., p. 124.)

Sur une comète vue à l'embouchure du fleuve de Mississipi en février et mars. (Mém. acad., p. 223.)
1703.—Observations de l'éclipse de lune du 3 janvier faites à Paris et à Rome, comparées ensemble. (Mém. acad., p. 5 et 23.)
Les observations de l'équinoxe du printems de cette armée 1703, comparées avec les plus anciennes. (Mém. acad., p. 41.)
Table où les quatorzièmes pascales sont distribuées dans le cycle de 19 ans, suivant le concile de Nicée. (Hist. acad., p. 91.)
Sur une conjonction de Jupiter avec Saturne. (Hist. acad., p. 89.)
1704.—Réflexions sur des Mémoires touchant la correction grégorienne communiqués par M. Bianchini. (Mém. acad., p. 142.)
Des équations des mois lunaires et des années solaires. (Mém. acad., p. 146.)
Observation de l'éclipse de lune du 17 juin. (Mém. acad., p. 197.)
Occultation de Jupiter par la lune, observée en plein jour. (Mém. acad., p. 233.)
Conjonction de Jupiter avec la lune le 24 août. (Mém, acad., p. 247.)
Observation de l'éclipse de lune du 11 décembre. (Mém, acad., p. 356.)
1705.—Il est le premier qui ait employé les éclipses de soleil à la recherche des longitudes. (Hist. acad., p. 122.)
Réflexions sur les observations des satellites de Saturne et de son anneau. (Mém. acad., p. 14.)
1706.—Réflexions sur les observations envoyées à M. le comte de Pontchartrain, par le père Laval, sur les réfractions astronomiques. (Mém. acad., p. 78.)
Observations d'une comète qui a commencé mars. (Mém. acad., p. 91 et 148.) à paraître au mois de Sur les taches du soleil. (Hist. acad., p. 121.)
Observation de l'éclipse de lune du 28 avril. (Mém. acad., p. 155.)
Observation de l'éclipse de soleil du 12 mai, et réflexions. (Mém. acad., p. 169 et 249.)
Recherches sur la parallaxe de Mars. (Hist. acad., p. 99.)
1707.—Observation de l'éclipsé de lune du 17 avril au matin. (Mém. acad., p. 168.)
De la dernière conjonction écliptique de Mercure avec le soleil. (Mém. acad., p. 175.)
Des irrégularités de l'abaissement apparent de l'horizon de la mer. (Mém. acad., p. 195.)
Réflexions sur les observations de Mercure. (Mém. acad., pag. 85 et 359.)
Observations d'une comète. (Mém. acad., p. 558.)
1708.—Réflexions sur la comète qui a paru vers la fin de l'année 1707. (Mém. acad., p. 89.)
Observation d'une comète qui a paru à la fin de novembre 1707, faite dans l'Observatoire du comte Marsigli. (Mém. acad., p. 323.)
Observation de l'éclipsé de Vénus par la lune du 23 février. (Mém. acad., p. 106.)
Observation de l'éclipsé de lune du 5 avril. (Mém. acad., p. 182.)

Observation du passage de la lune par les étoiles méridionales des pléiades, le 10 août au matin (Mém. acad., p. 407.)
Observations de l'éclipsé de lune du 29 septembre, faites à Paris, à Marseille et à Gênes. (Mém. acad., p. 404 et 418.)
Réflexions sur les éclipses du soleil et de la lune du mois de septembre. (Mém. acad., p. 410 et 412.)
Globe céleste construit par rapport au mouvement des étoiles fixes. (Mém. acad., p. 97.)
1709 (1).—Du mouvement apparent des planètes à l'égard de la terre. (Mém. acad., p. 247.) [(1) Une éclipse de soleil, observée cette année, fit voir dans les lieux où elle fut totale cette chevelure lumineuse autour de cet astre que J.D. Cassini avait prédite comme devant avoir lieu, selon son hypothèse sur la nature de la lumière zodiacale.]
1711.—Il devient totalement aveugle.
1712.—Il meurt le 14 septembre, âgé de 87 ans et 3 mois.

Il a laissé une grande quantité de manuscrits et de traités astronomiques; plusieurs pièces de vers latins et italiens, entr'autres une cosmographie ou description du monde très-étendue en vers italiens, dont on a rapporté un fragment; une correspondance considerable avec les savans de son tems; des tables du mouvement du soleil et de la lune; un ouvrage, qui n'a pas été publié, sous le titre de: Magna Periodus luni-solaris et pascalis, duobus libris comprehensa, ect… Il existe encore dans les journaux d'Italie, dans des receuils, et dans plusieurs ouvrages particuliers, des lettres, des expériences et différens Mémoires de Cassini, que l'on n'a pu citer ici. Ce qu'on en a rapporté suffit pour donner une idée de l'immensité de ses travaux. Il est peu de partie de l'astronomie qu'il n'ait ou ébauchée, ou étendue, ou enrichie de quelque découverte. C'est à lui que l'on doit les théories et la première détermination exacte des réfractions et des parallaxes; la théorie et les premières tables exactes du soleil et du mouvement des satellites de Jupiter et de Saturne. Il a eu la plus grande part à la détermination des longitudes terrestres, rendue d'un usage presque journalier et universel par ses éphémérides des éclipses des satellites de Jupiter; la méthode de déterminer les mêmes longitudes par les éclipses de soleil est due à lui seul. Il a découvert quatre satellites de Saturne, la duplication de son anneau, la lumière zodiacale, les taches sur le disque des planètes, et celles des satellites de Jupiter, la rotation des planètes, dont il a déterminé le tems de la révolution sur leurs axes. On lui est encore redevable de la solution des plus importans problèmes de l'astronomie, des méthodes et des explications les plus ingénieuses. Enfin l'on peut dire que la perfection de l'astronomie et des nouveaux instrumens depuis cent ans, n'a changé que peu de choses à plusieurs des déterminations fixées par J.D. Cassini.

Bibliography

Books and Articles

Jean-Dominique Cassini, Mémoires pour servir à l'histoire des sciences et à celle de l'Observatoire de Paris, Parigi, Bleuet, 1810
Anna Cassini, Gio: Domenico Cassini – Uno scienziato del Seicento, Comune di Perinaldo 2003
Anna Cassini, I Maraldi di Perinaldo, Comune di Perinaldo 2004
A cura di Stellaria, Il Museo G. D. Cassini, Comune di Perinaldo 2006
Van Helden, Albert "Contrasting careers in astronomy: Huygens and Cassini", De zeventiende eeuw. Jaargang 12 (1996)
Jacques Cassini, Traité de la grandeur et de la figure de la terre, 1723. pp. 182–3 & p. 302
Gabriella Bernardi, Il borgo delle Stelle, L'Astronomia, 05 14 2008, pp. 48–55
Gabriella Bernardi, Il Gesuita che portò l'astronomia in Cina, Le Stelle n. 133, August 2014, pp. 61–65
Gabriella Bernardi, The Unforgotten Sister Female Astronomers and Scientists before Caroline Herschel, Springer 2016
Devic, M.J.F., Histoire de la vie et des travaux scientifiques de J.D. Cassini IV, Clermont 1851
René Taton and Curtis Wilson, The General History of Astronomy—Planetary astronomy from the Renaissance to the rise of astrophysics, Cambridge University Press 1989
Dava Sobel, Longitudine. Come un genio solitario cambiò la storia della navigazione, Rizzoli 1999
Matt Williams, Who was Giovanni Cassini?, Universe Today, 30 November 2016
Pedersen Kurt M., Une mission astronomique de Jean Picard: le voyage d'Uraniborg, in Jean Picard et les débuts de l'astronomie de precision au XVII siècle, Actes du Colloque du tricentenaire, édités par Guy Picolet, C.N.R.S., 1997
Marta Cavazza, La cometa del 1680–1681: astrologi e astronomi a confronto, Studi e Memorie per la Storia dell'Università di Bologna, Nuova Serie, vol. III, 1983
Jean-Baptiste Joseph Delambre, Histoire de l'astronomie au dix-huitième siècle, édité par Claude-Louis Mathieu, Bachelier, Paris, 1827
Laurence Bobis, James Lequeux, Cassini, Rømer and the velocity of light, J. Astron. Hist. Heritage. 11 (2): 97–105, 2008
Francesco Savorgnan di Brazzà, L'opera del genio italiano all'estero, La Libreria dello Stato, 1941

Web

http://catnaps.org/cassini/family.html
http://www.treccani.it/enciclopedia/cassini/

G. Bernardi, *Giovanni Domenico Cassini*, Springer Biographies,
DOI 10.1007/978-3-319-63468-5

http://www.treccani.it/enciclopedia/giovanni-domenico-cassini_(Il-Contributo-italiano-alla-storia-del-Pensiero:-Scienze)/
https://en.wikisource.org/wiki/1911_Encyclop%C3%A6dia_Britannica/Cassini
http://www.universetoday.com/130823/who-was-giovanni-cassini/
https://fr.wikipedia.org/wiki/Acad%C3%A9mie_des_sciences_(France)#Historique
https://en.wikipedia.org/wiki/French_Academy_of_Sciences#See_also
http://www.academie-sciences.fr/en/Histoire-de-l-Academie-des-sciences/history-of-the-french-academie-des-sciences.html
http://rumsey.geogarage.com/maps/cassinige.html
https://www.seeker.com/star-trek-inspiration-meet-the-real-jean-picard-1765425621.html
https://www.leidenuniv.nl/fsw/verduin/stathist/huygens/acad1666/lunardraw.htm
https://www.nasa.gov/mission_pages/cassini/spacecraft/index.html
http://www.esa.int/Our_Activities/Space_Science/Cassini-Huygens